BACKYARD FARMING

Make your home a homestead

RAISING GOATS

For Dairy and Meat

Kim Pezza

hatherleigh

hatherleigh

Hatherleigh Press is committed to preserving and protecting the natural resources of the earth. Environmentally responsible and sustainable practices are embraced within the company's mission statement.

Visit us at www.hatherleighpress.com and register online for free offers, discounts, special events, and more.

Backyard Farming: Raising Goats

Library of Congress Cataloging-in-Publication Data is available upon request.
ISBN: 978-1-57826-473-5

Cover Design by Carolyn Kasper
Interior Design by Nick Macagnone

Printed in the United States
10 9 8 7 6 5 4 3 2 1

TABLE OF CONTENTS

INTRODUCTION

Some of the most fun you'll have as a new homesteader or backyard farmer is your time spent with goats. In short order you'll see that there is a reason why these magical animals are known as the "clowns of the barnyard." In fact, I've had a few that exhibited shades of Houdini as well. It is all part of the fun and frustration of owning goats.

Goats like to sleep in the strangest places...

Goats are available in a number of colors, sizes, varieties, and (as you will find out) personalities. They can be used as dairy, meat, working, or pet animals. They can even be used as lawnmowers in places where mowing may be difficult (although weed killers cannot be used for areas where the goats crop and graze).

Goats are a great addition to the farm, especially if you want milk or meat but don't have the space for cattle or the ability or patience to deal with large animals. They are even beginning to make an appearance on urban and city farms. Best of all, goats are animals that the entire family can safely handle and work with.

So start reading and start planning; that new goat is closer than you think.

MEET THE EXPERT

Kim Pezza grew up among orchards and dairy and beef farms having lived most of her life in the Finger Lakes region of New York state. She has raised pigs, poultry and game birds, rabbits and goats, and is experienced in growing herbs and vegetables. In her spare time, Kim also teaches workshops in a variety of areas, from art and simple computers for seniors, to making herb butter, oils, and vinegars. She continues to learn new techniques and skills and is currently looking to turn her grandparents' 1800s farm into a small, working homestead.

CHAPTER 1

A BRIEF HISTORY OF THE GOAT

The goat is one of the earliest, if not *the* earliest, example of an animal to be domesticated by humans. The word *goat* actually stems from the old English *gat,* meaning "she-goat," whereas the word for male goat was *bucca,* which we know today as "buck." By the eighteenth century, females would also come to be known as "nannies" or "nanny goats," while males would become known as "billys" in the nineteenth century.

A member of the *Bovidae* family, the goat was domesticated in the Zagros Mountains of Iran. While our modern goat breeds are subspecies of the wild goats found in southwest Asia and eastern Europe, it is thought that the **wild bezoar** (wild Iranian goat) is the origin of all domestic goats.

Remains of goats have been found that date back as far as 10,000 years at sites including Jericho in Palestine and Djeitun in what is now Afghanistan. They seem to have been usually kept in herds under the watchful eye of children or teens, referred to as shepherds.

For better or worse (from the goat's point of view), goats played a role in a number of histories, mythologies, and other belief systems. The Greek satyr Pan was half man and half goat, with the head and torso of a man, and the ears, horns, and lower body of a goat. Although he was a god of fields, words, and flocks, Pan did not always carry a good reputation with him; it is from Pan that we derive the words *panic* and *pandemonium.*

In ancient Syria, records relate the story of a she-goat, which was draped in silver necklaces and driven away from the town in a ritual of purification of the town in anticipation of the king's wedding. It was thought that the animal would also carry away any and all evils surrounding the town and that day. And before the Yule goat was made of straw, the title referred to the goats slaughtered at the festival of Yuletide. In Scandinavia, the Yule goat is seen as the deliverance of gifts and glad tidings. However, the Finnish see the Yule goat as something more horrific, something that scares away evil and bad luck.

By the sixteenth century, goats had arrived in North America with the Spanish explorers, with the English bringing them to New England in the seventeenth century. However, it is also said that Columbus brought goats with him when he arrived in America in 1492.

Although goats as a whole did not have much value until the Civil War, by the 1850s Angora goats were being imported to the United States for their hair, which is known as **mohair.** The main production area for mohair was Texas.

The advent of the twentieth century witnessed the introduction of European dairy breeds into the United States, with breeds such as Saanen, Alpine, and Toggenburg, which would soon replace or be crossbred with English and Spanish goats due to their outstanding production. The first pure meat goat, the Boer, was imported to the United States in 1993 from South Africa.

Considered small livestock, there are over 400 million goats worldwide still being used for milk, meat, work, hide, and hair as they were centuries ago. In fact, more goat meat and milk is consumed worldwide than that of any other animal.

However, regardless of how long goats have been domesticated and (as of the twentieth century) looked upon as pets for some households, if left to their own devices, goats can easily become feral (wild) animals, showing just how tough and self-reliant these animals still can be.

Goat cheese on pita bread is a delicious example of the goat's contribution to cuisine. Photo by jeffreyw under the Creative Commons Attribution License 2.0.

This book will guide the potential new goat owner through the basics of keeping these highly useful and ever-comical creatures on his/her homestead or urban farm (yes, some cities *do* now allow goats). It will give you an idea of the different aspects of care providing and help in your decision to bring goats into your home. Whether kept for milk, meat, or both, a goat is one of the best investments a new homesteader or urban farmer can make if they are in a position to do so.

As a final note, you will see the following terms used throughout, which refer to the various stages of a goat's life and its sex:

- **Female:** doe, nanny
- **Intact male:** buck, billy
- **Castrated male:** wether
- **Offspring:** kids

CHAPTER 2

TYPES AND BREEDS

There are basically two types of goats: **dairy goats** and **meat goats.** Despite this, there are still over 300 breeds to choose from, running the gamut in terms of appearance. This includes a wide variety of **heritage breeds:** previously popular breeds that were replaced by more commercially viable varieties, but which are still well suited to small farm production.

Depending on the breed, a goat may stand as short as 16 inches to as tall as 35 inches. Weight-wise, they can run from 22 pounds to 300 pounds for does and 27 pounds to 350 pounds for bucks, while life expectancy averages 8 to 12 years, although some animals have lived longer.

Many owners will choose to remove wattles.

Both does and bucks can have horns, beards (the common term for the hair coming down from the chin, just like a beard), and wattles. Wattles are growths of hair-covered skin that hang on either side of the neck of the goat. There are usually two, but there can be only one. As of yet, no specific reason has been found for a goat to have wattles, so some owners will choose to remove them when the goat is young, usually by cleanly snipping them off.

Unlike cattle, whose horns are actually made of **keratin** (like a human fingernail) with a bone center, the horn of the goat is composed entirely of live bone. Also unlike cattle, both bucks and does have horns. Although this will be discussed further in Chapter 6, it is common practice among goat owners to disbud (removing the little horn bumps) from dairy goats at a very, very early age. However, some goat owners are getting away from this practice. The removal of the horn is called **polling.** And while some animals, like cattle, may be born polled, this is rare in goats. This would also be a good time to mention that goats may have accidents in which a horn (or two) can be broken. If it looks like an injured or damaged horn will need attention, contact your veterinarian, as you are dealing with live bone.

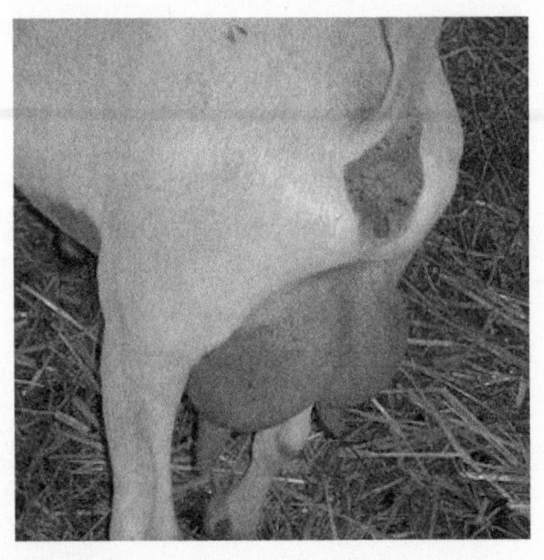

A goat's udders are the source of milk, both for its kids and its owners.

Does will also have **udders,** just like a dairy cow, save for the fact that a goat will have two teats. Like the cow, the udder holds the milk for the kid and, in the case of the dairy goat, for milking.

These traits run true through all the goat breeds; however, meat-specific types will not have the exact same traits in the area of the udder, as meat-specifics are bred for meat and not for milking. Let's now look at the various types of goats individually, along with examples of each breed.

Dairy

Dairy goats are built and bred for milk production. Virtually any breed of dairy goat will supply you with all the milk you'll need; therefore, when selecting your goats, choose not only for your needs, but also for your likes. In other words, pick a breed or individual animal that appeals to you. You will be working with these animals daily, so it is important that you enjoy what you have. You should also keep in mind that in order for does to produce milk, they usually must be bred yearly. So you will either need enough room to keep the future kids or have the foresight to plan out what you are going to do with the kids once they are ready to leave.

In the following descriptions, the term *butterfat* will be used. **Butterfat,** also known as *milk fat,* is the fatty part of the milk that gives goat milk its sweet flavor and also affects the texture. It is additionally the chief component in goat's milk butter. You will also notice milk being discussed in pounds (weight), not liquid measurements. This is for accuracy and ease in keeping milk records. Keep in mind that approximately 8.2 to 8.5 pounds of milk equals one gallon.

In the United States, the most common dairy breeds are:

- Toggenburg
- Nubian
- La Mancha
- Saanen
- Sable
- Oberhasli (heritage breed)
- Alpine
- Nigerian Dwarf (heritage breed)

Toggenburg

The oldest known breed of dairy goat, the **Toggenburg** was developed 300 years ago, originating in the Toggenburg Valley of Switzerland. A medium-size goat, the bucks range from 150 to 200 pounds and 34 to 38 inches high, while does run 125 pounds and up and 30 to 32 inches high. Although it is not uncommon to be able to milk the Toggenburg doe without having to rebreed each year (up to twenty months is not uncommon), its milk is low in butterfat, usually having only 2 to 3 percent.

Toggenburgs also make good pack and work animals (castrated males or wethers usually provide this service).

Probably one of the most famous Toggenburg owners and advocates was Lillian Sandburg, wife of poet Carl Sandburg.

Nubian

Also known as **Anglo Nubian,** this is an English breed developed during the time of the British empire. They are a cross between the old English Milch and other breeds, including Nubian bucks imported from Russia, India, and Egypt. They have long hanging ears and a convex nose.

The Nubian is a larger breed than the Toggenburg, with Nubian does running 135 pounds and 30 inches at the withers (which is the ridge between the shoulder blades) and bucks running 175 pounds and 32 inches at the withers. The Nubian is also fleshier than other dairy breeds, making the Nubian a good dual-purpose (milk and meat) goat as well.

Nubians are very intelligent, which unfortunately enables them to get into mischief easily. However, if you treat them well and give them proper food, water, housing, and attention, they will tend to behave.

Nubian milk is high in butterfat, having about 5 percent butterfat content. However, their overall milk production is lower than average. For those that live in cooler climates, the Nubian is good to 0°F.

La Mancha

The **La Mancha,** also commonly known as the *American La Mancha, i*s the only breed developed in the United States, although it is not certain exactly which breeds were crossed to create the animal. The main trait that makes the La Mancha stand out is that the goat appears earless due to the ear flap being so tiny. This is also known as a *gopher ear.*

The La Mancha is a docile goat and has good udders. They are high-quantity milk producers, with does producing up to 1.5 gallons per day, depending on breeding. The milk is fairly high in butterfat at about 4 percent. Size-wise, does run a minimum of twenty-eight inches at the withers and around 130 pounds, while bucks run at least thirty inches and around 155 pounds.

Saanen

The **Saanen** is the largest of the dairy goats, with does running upward of 150 pounds and bucks 200 pounds and up. Originating in the Saanen Valley in the south of Switzerland, the breed was imported to the United States sometime in the early twentieth century, after which subsequent importations came from England.

The Saanen is also called the "Holstein" of the goat world, with does producing about a gallon of milk per day, although their butterfat is low at only 2.5 to 3 percent. However, due to its excellent rate of milk production, calm personality, ease of keeping, and ability to adapt to its environment, the Saanen is a preferred breed for commercial dairies. Both males and females are white.

Sable

The **Sable,** or *Sable Saanen,* is basically a Saanen that is not white (which is caused by recessive genes). Colors include black, brown, gray, and some white, although a Sable cannot be totally white. Most traits, including those governing their value as dairy goats, are the same as Saanens; however their color makes them better suited for tropical climates than their white counterparts, which can contract skin cancers due to their light skin. The Sable was not recognized as a breed in its own right until 2005.

Oberhasli

Also known as the *Swiss Alpine,* the **Oberhasli** is a very old breed and originated from the canton of Bern in Switzerland (the same as the Saanen).

The Oberhasli is known as a **color breed.** In order to conform properly to breed, the Oberhasli must be correctly colored in order to be registered. The color is called *chamoisee,* with the breed having a red bay with black head, legs, stomach, and dorsal (although does—and only does—may be all black).

The Oberhasli was first imported to the United States in 1906. Unfortunately the line was lost due to crossbreeding. However, in 1936, a new, pure line of four does and one buck was imported to the United States. As a result, all the Oberhasli in the United States today can be traced back to those five animals. The breed is still considered rare in the United States.

The Oberhasli is a decent-size goat, with does running 28 to 30 inches at the withers and 120 to 150 pounds, while bucks run 30 to 32 inches at the withers and 150 to 175 pounds.

The Oberhasli is a producer of sweet milk, with does averaging six to eight pounds of milk per day (up to a gallon) with 3.4 to 4 percent butterfat.

Alpine

Also known as the **French Alpine,** this goat, as with all of the other European mountain goats, is thought to be a descendent of the Pashang or Bezoar Goat. A medium breed, the Alpine doe is approximately 30 inches at the withers and 135 pounds, while bucks will stand 34 to 40 inches at the withers and 170 pounds.

The Alpine is known for being a good milker, producing up to and over four pounds of milk per day at 3.5 percent butterfat. Due to its heavy milking and the fact that it is hearty and adaptable, the Alpine (like the Saanen) is a favorite commercial breed.

Nigerian Dwarf

The **Nigerian Dwarf** is not only a dairy goat, but also a dwarf goat as well. Originating in west Africa, it is believed these little goats were originally brought on board ships to provide food for big exotic cats being shipped to the United States, with those goats that survived the trip going to live in zoos.

A perfectly proportioned miniature dairy goat, the Nigerian Dwarf comes in many colors. Recently popular as pets, they are gentle and trainable. As far as size, there are two schools of thought on this. The Nigerian Dwarf Goat Association states that the does should be 17 to 19 inches (21 inches at the most) at the withers, while bucks should be nineteen to 21 inches (and no more than 21 inches) at the withers. However, the American Goat Society says that the does should be less than 22 inches at the withers and bucks less than 23.5 inches at the withers.

The average weight of a Nigerian Dwarf is about seventy-five pounds. Their size makes them great for the urban, suburban, or city farm. Despite their small stature, Nigerian Dwarfs are excellent milk producers, producing anywhere from 1 to 8 pounds of milk per day at anywhere between 6 to 10 percent butterfat.

There are a few things besides size that differentiate the Nigerian Dwarf from its full-size counterparts. First, Nigerian Dwarf goats are not disbudded. In fact, having no horns goes against the confirmation requirements of the breed and will disqualify a goat from a show. And unlike the larger dairy breeds, the Nigerian Dwarf may be bred during any season, allowing the goat owner to stagger breeding so he/she never has to be without goats for milking.

It is worth noting that crossbreeding with the Nigerian Dwarf has also been used to "miniaturize" other dairy breeds.

Meat

Although any goat may be used for meat (even dwarf breeds, although they don't produce a lot of meat), those goats bred specifically for meat use will be larger built and also bred for muscle and carcass development. In fact, many were dairy-type breeds, selectively bred to produce bigger, meatier, and/or faster-growing animals specifically for meat use.

Some of the more popular breeds (either meat specific or bred up for meat use) are:

- Boer (South African)
- Spanish Meat Goat
- Tennessee Meat Goat
- African Pygmy

Boer Goat

Also called the *South African Boer* due to its South African roots, the **Boer** was developed specifically for meat use. Developed in South Africa in the 1900s, the breed was not introduced into the United States until 1993 to try to meet the demands of the growing goat-meat market. Its high growth rate and excellent muscle and carcass condition, along with excellent fertility, have made the Boer one of the most well-known meat breeds in the United States and world.

For the most part, bucks have very distinctive coloring, usually solid white with red or black heads. However, they may also be found in solid black, solid red, or pinto. Bucks can reach weights between 250 and 300 pounds, while does weigh anywhere from 150 to 250 pounds. Height-wise, it is considered a large breed, about the same height as the Saanen or Nubian dairy breeds. The Boers have a similar look to that of the Alpines, with large ears and convex noses; however, they have larger frames and are heavier boned.

A Boer goat. Photo by just chaos under the Creative Commons Attribution License 2.0.

It is interesting to note that the milk of the Boer is quite high in butterfat. However, due to the fact that their short legs make milking difficult and there is not usually enough milk left to bother with after the kids have finished, they have not yet been really considered for milking. Certain classes of Boers and dairy goats, such as Nubians, though, have gained popularity for producing a dual-purpose type of goat.

Spanish Meat Goat

The **Spanish Meat Goat** is a descendent of the Spanish Goat, brought to the United States by the early Spanish settlers in the 1500s. The Spanish Meat Goat was later developed through selective breeding to produce the biggest and meatiest bucks possible.

The Spanish Goat/Spanish Meat Goat is also known as a brush or scrub goat, as it was used to clear away scrub. They are also known as briar goats in North and South Carolina, wood goats in Florida, and hill goats in Virginia.

From the 1800s onward, the Spanish Meat Goat was used by some for meat, but its primary use was in clearing scrub. This was true until the late 1980s, when the United States market for goat meat was growing, at which point the scrub-eating goats became meat goats. The Spanish Meat and Tennessee Meat Goats were then the only known meat goats in the United States until the introduction of the Boer Goat in the 1990s.

Eventually, ranchers and farmers began crossing the Spanish with the Boer. However, with each successive generation when the new offspring were bred, more and more Boer characteristics were appearing, while the hardiness and almost hands-off nature of the Spanish were disappearing. Not wanting to lose these characteristics that had made the Spanish so appealing to keep and raise, these farmers and ranchers went back to the drawing board and started to breed back by crossing Spanish does with Boer bucks.

Spanish Meat Goats have also been crossbred with Nubians, producing a better-size animal with improved milk production and meatier kids. However, this crossbreeding, as well as the changeover to Boers, led the Spanish Goat (the only goat known in the United States for 300 years) to become an endangered livestock species. The Spanish is now on the American Livestock Breeds Conservancy's conservation priority list.

Spanish Meat Goat bucks can run up to 250 pounds, while does run 100 to 125 pounds.

Tennessee Meat Goat

The **Tennessee Meat Goat** was actually originally produced from the Tennessee Fainting Goat, a breed known for its "fainting" trait when startled. In point of fact, fainting goats (of which there is only one breed) are **myotonic**, a condition that causes the muscles to stiffen and lock up, which in turn causes the animals to fall over, which is why they are also known as the *Wooden Leg Goat.*

And although one would think this would be a detriment to the Tennessee Meat Goat, it is said that the myotonic condition may result in more tender meat and a higher ratio of meat to bone.

The Tennessee Meat Goat came about through the selective breeding of the Tennessee Fainting Goat, with farmers invariably selecting the larger, meatier animals for rebreeding.

The Tennessee Meat Goat is actually smaller than the average meat breed, having a wide body and short legs, but the breed is particularly meaty. The goat comes in a variety of colors, including black and white, all white, black, tan and white, and tan and roan.

The Tennessee Meat Goat is a hearty, fertile goat with a long breeding season. And as the goat is crossbred with another type (such as a Boer), the fainting or myotonic condition usually does not come up with the offspring.

African Pygmy

The **African Pygmy Goat** is listed as a meat goat, although it is, for the most part, kept as a pet in the United States. However, regardless of whether it is being used for its meat or just as a pet, the African Pygmy is a very popular goat to own.

The African Pygmy Goat originated in the Cameroon Valley of west Africa, and was then imported to the United States from Europe, mostly for zoos and research. However, once these little goats were obtained by private breeders, they quickly became popular as pets.

The African Pygmy comes in a variety of colors. They are hardy and can easily adapt to their surroundings. They are friendly goats and, with patience, are trainable (even litter-box trainable). However, they can also be quite mischievous.

Because of their small stature, does can be anywhere from 53 to 75 pounds, and bucks can range from 60 to 85 pounds, with the heights for both at 16 to 23 inches at the withers. They are also a good choice for the urban, suburban, or city farm.

Unlike the full-size meat breeds, Pygmies can be milked, although their milk production isn't like the Nigerian Dwarfs', so they will produce less milk per milking. Also important to keep in mind is that the African Pygmy likes companionship, although the companion does not have to be another goat. My Pygmy Goat's best friends are my Australian Shepherd and my blue-and-gold macaw. And as these small goats are prey animals, they need to be protected from predators, especially at night.

Fiber

A third use for which people raise goats is for their hair or fiber. The most well known of the fiber producing goats are the Angoras. First introduced into the United States in 1849, the fiber the Angora produces is known as mohair. The goats are shorn twice a year. In the United States, Texas is the major producer of mohair, and its production places third in the world.

Of course, there are many more types of goats in the world used for both meat and milk. This chapter covers only the most popular varieties and includes a few heritage breeds as well. When trying to select your goats, read about all the breeds that you are interested in. Talk to breeders and visit some farms if you can. Choose the breed best suited for your needs and situation. And although some, like the Angora, need some special care, basic care is the same for all goats. Hooves may need to be trimmed if there is nothing to help them naturally wear down, as well as an occasional ear cleaning and burr removal. As stated before, choose a breed or breeds that you will enjoy, as it will make owning the goats much more fun for both you and them!

CHAPTER 3

HOUSING

The type of housing you will need to provide for your goats will depend on what type of goat you have and the climate you (and they) will live in. Housing can be as simple as a fenced-in barnyard with a small three-sided shed or as big as a pasture with a barn. Whatever you use, it must be clean, dry, draft free, and have good ventilation.

Fencing is another area of importance for your goat. As goats can be very adept at finding their way over fences (or through them), you will need to make sure that your fencing is not only predator proof, but also goat proof as well. Let us now take a quick look at housing your goats.

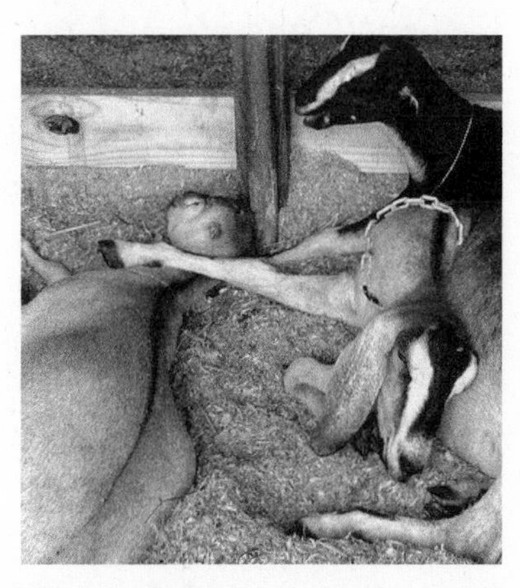

While goats will often sleep close together, especially during times of cold or rain, you still need to ensure they have adequate space.

So long as it is **clean, dry,** and **draft-free,** housing for your goats can be quite flexible, and fit any range of budgets. When deciding on what material to use or what to build for goat housing, keep in mind that you will need to clean your goats' shelter regularly; be sure it will be easy to keep clean. Also, make sure you allow plenty of room for your goats; **fifteen to twenty-five square feet per animal** is the minimum recommended amount of space for housing. Of course, the amount of space your goats will require will also depend on the type of goat you have, as, obviously, a pygmy will need less space than a Saanen. Don't let the way your goats sleep make you believe they can do with less than the minimal space. Although goats will sleep together in little groups, they still need space. You will find your goats will spend time indoors during inclement weather (especially heavy rain), on hot days (if the shelter is cool), or on the rare occasion when one or two would just rather stay indoors.

Flooring

You will also need to decide on the type of floor that your goat housing will have. Theoretically, you can have any flooring of your choosing. In reality, there are differences in the types of floors in terms of ease of cleaning and ease on your goats.

Cement/Concrete

While a cement floor may seem to be the best choice, especially when it comes to cleaning, it can have definite disadvantages. Even when covered with straw, cement can hold the urine smell, and when you go to clean it, you may have quite a wet floor. Except for absorption by the straw, the urine will have nowhere to go. Also, and especially in cold northern climates, a concrete floor can be awfully cold and hard on the goats, even when packed with straw.

Wood

If you're using an older preexisting shed on your property, it may already have a wood floor, depending on the original purpose. While a wooden floor will, no doubt, be more comfortable for your goats than cement, you could end up with the floors rotting out, or the wood could hold the urine scent. You will still need to pack the floor with straw, and if you allow the floor to get too wet from urine-soaked straw and the urine itself, you can begin to have problems with rot.

Dirt

A dirt floor is actually the best choice for your goat housing. It stays warmer than cement for your goats, and the dirt will absorb the urine. Again, it must be covered with straw, but you won't face rot, and cleanup will remain reasonable; bedding removal of a dirt floor is quite easy.

It should be noted that **straw**, not hay, is what should be used for goat bedding. Unlike hay (which will be covered in Chapter 4), straw is hollow; grain stems, such as wheat, barley, or oats, make a good, warm bedding for goats and most other animals. Straw is also golden in color. Although your animals may eat it (and it won't hurt them if they do), straw lacks both nutrition and taste.

Buildings

Your goats' housing should give them **easy access to the outdoors,** but also the ability for you to lock them in at night if need be. Of course, if you already have a good-size barn, then your housing problem is nonexistent, especially if you will be keeping multiple goats. But if you don't have any satisfactory buildings, then you will need to build something that is big enough for your animals, easy to clean, and will keep your goats comfortable throughout the seasons.

Although a barn is nice, not everyone can afford to put one up. And if you're only going to keep a few goats, a simple three-and-a-half-sided shed will work fine. Most would say a good, dry, deep **three-sided shed** is fine. However, the extra half side grants your animals further protection from wind and the elements. The floor should be dirt and deep with straw, keeping it dry and draft free.

If you are breeding or milking, you will want more protection, especially if you are breeding and/or milking a number of goats. When keeping a large number of goats, some type of large barn or even a few small barns (or buildings) are really the best fit. With large buildings, you can have stalls, feeding areas (for the times when you need to feed indoors), and, for dairy goats, the necessary milking parlor (which we will discuss in Chapter 7).

Building a goat shed provides your goats with adequate, easy-to-maintain living space.

Stalls are especially useful for does near kidding and does with very young or newborn kids, especially those born during harsh winters. As goats are prey animals, baby goats (especially newborns) are even more vulnerable. Basically, it is nice to be able to have a "me space" for the does and their kids for the first few weeks or so.

Penning your goats provides you with the useful ability to separate goats, which can be valuable during kidding time.

Stalls are also nice to have if you need to isolate a goat for any reason, especially for injury, when being with the rest of the herd could be dangerous. Alternatively, you may need to isolate a goat for an illness, both to treat the animal and to keep it segregated from the rest of the herd, or if introducing a new goat into the herd. And if you are in an area prone to blizzards or other weather situations where your goats really shouldn't be outside, an appointed feeding spot in the building is quite useful.

Other options for goat housing would be old garden sheds or

chicken coops (that are well cleaned and sterilized). Some people have even transformed old camping trailers or garages (provided they no longer hold vehicles). As long as you remember that your goats' home needs to be clean, dry, and draft free (those three key words), any adequate amount of space can be transformed into a home for your goats; you can be as simple or creative as you want.

Fencing

Along with housing, you will also need fencing to keep your goats contained. Note that keeping goats contained can be a feat in itself for the new goat owner and even for some experienced owners. If your fence is too short, then they'll be over it in a second. Too weak, and they will knock it right down and walk over it. Too big a space in the wires, and they will poke their heads through so much you'll end up with huge holes that they will just continue to make bigger and bigger until they can finally jump right through. The best fencing for goats is a heavy-gauge, woven fence with small openings in the wire and climb proof, usually known as **goat fencing.** Although chicken wire does, indeed, have small openings, even the heaviest gauge wire will not be strong enough for your goats' destructive tendencies. Barbed wire should *never* be used with your goats, as it can be very dangerous. Some owners may think it a good idea to string a single strand of barbed wire across the top of the regular fence. However, when your goats successfully make it over the fence (and at least one will), the barbed wire can scratch their faces and eyes, tear their hides, or damage their udders.

Depending on what type of fencing you use, an electric fence can work with goats. You will have to use more than a single strand to keep them in, as, unlike horses (who will stay contained with only a few wires strung around the pasture), goats will most likely find a way around a fence if it consists of only a couple wires, electricity and all. A combination of goat fencing and a few strands of electric fencing can work well together, if you find that you do need electric fencing at all.

Climb-proof fencing is basically woven so the openings at the bottom of the fence are too small for a goat to get a good foothold on. Of course, in your goat career, you may still have one or two that figure it out. As the fence gets higher, the weave openings get larger, but the areas most likely to allow a goat to climb have been "goat proofed." Some may choose to run a single electric-fence wire across the top of the regular fence for extra predator protection. This could help and certainly won't hurt, if you think you may have problems. (A coyote can climb a six-foot-fence, so keep this in mind when deciding on height.)

When installing your fence, allow extra height for burying a bit of the fence. This is due to the fact that, if they find a loose spot or space under the fence, the goats will work at it until they can successfully crawl under. However, please note that if your fence is heavy enough and the posts are put in close enough together, then you will not need to worry about sinking the fence. If you are worried about your goats digging out under the fence, don't. Goats will not dig under a fence. But this does not mean that a predator may not. To help solve this problem, metal guards, which are long, narrow sheets of galvanized metal, are sunk into the ground along the fence line. They do work well. Other goat owners will build guards along the fence line (inside, if you're concerned about animals digging out; outside, if you're afraid of animals digging in). This consists of a long, narrow piece of chain link that lays flat in a shallow trench along the fence line, and that is then covered with soil. Animals cannot break through the chain link.

But before you can put your fencing up, you need fence post. You can find posts in various sizes, as well as made from an assortment of materials, including steel and wood. The type of post used is basically your choice. The metal or **"T" posts** are metal posts that look like an upside-down T. They are a bit easier to install and less expensive than wooden posts, but traditional wood posts are aesthetically better and will give you the traditional pasture look, if that is important to you. You can also use a combination of wood posts and "T" posts, using wood posts in all corners and spaced about every 40 feet. You would then place the "T" stakes in between the wood posts every 10 feet or so.

When using wooden fence posts, you may be tempted to use **treated posts** to help protect them against weather and rot. But, if your goats end up with a fondness for chewing on wood, treated posts could be toxic to them. If you are afraid of rot, especially for the part of the posts that is buried, it is best to either purchase untreated posts, and then paint or treat only that part of the posts that will be buried, or ask your fencing salesperson if there are any treated posts available that are safe and nontoxic. And, if you are fencing in a wooded area, you may even be able to use some of the existing trees as fence posts. Simply attach the fence to the tree. Keep in mind that the tree will eventually engulf the piece of fence attached to it, but this usually does no harm.

There are many little hints and tips on putting up fencing, and a lot will have to do with the type of fencing you've chosen, the poles or stakes you're using, and the landscape of the ground you're fencing. Your local farm stores, extension offices, or fencing companies should be able to assist you in any installation questions you may have. You also have the option to have your fencing professionally installed, and while professional installation can be expensive, it is a viable alternative, if necessary.

Finally, before you put your fence up, you need to figure out how big your goat yard needs to be. As with the housing, you need a minimum amount of space per animal. For your goat yard or pen, you should allow about **200 square feet per goat.** Notice that this number may be a little smaller for pygmies and dwarfs, but this is the average. If you are housing your goats in a pasture, however, these numbers are a bit different; that will be touched on in Chapter 4.

A final note: many breeders will keep their breeding bucks separate from, but in view of, the does. This is due to the odor than an intact buck has, as there exists the possibility of the odor tainting the milk. It is also to prevent unplanned kids and control breeding times.

CHAPTER 4

FEEDING

Just like cows and sheep, goats are ruminants, meaning they have a four chamber stomach. The chambers consist of:

- Rumen
- Reticulum
- Omasum
- Abomasum

You've probably seen the old cartoons of the nanny goat eating everything that she can get a hold of, including tin cans. And while goats will eat paper and chew on wood, maybe even cloth, a goat will not eat tin cans—that is just an old wives' tale. However, you *do* need to watch what they eat, especially to make sure that they do not overindulge on grains. You will also need to ensure that their pasture and yard are clear of poisonous plants and little things that they might pick up along the way.

But what does a goat eat, or rather, what is a goat *supposed* to eat, anyway?

The Ideal Diet for Your Goats

Contrary to popular belief, goats do *not* prefer to pasture and do not graze. They prefer to browse on shrubs, brush, poison ivy, suckers, leaves, weeds, and the like. It is even said that, if you drink the milk of goats that eat poison ivy, it will build up your immunity to poison ivy. Some people will agree; some will not. Goats will eat grass, although it is not their preference. This might also be a good time to mention that your goat will eat your pine trees and bark off other trees, as well as the saplings. So bear this in mind as you place your fencing or allow your goats the run of the property outside their regular fenced areas.

Due to the fact that goats prefer weeds, brush, and such, many owners will use their goats to clear brushy areas, as the animals

Dangerous Herbs and Plants

Whenever your goats are out, but especially in new browsing areas, you need to watch for plants that are **poisonous** to your goats (although goats are generally pretty good about keeping away from poisonous plants, accidents can always happen).

The severity of a poisoning will depend on how much of the plant was consumed and the type of plant it was, but it is still in the best interests of you and your goats to scan pastures, yards, and any other areas you may let your goats loose in for these plants. Please note that this is *not* a complete list of toxic and/or deadly plants; these are just some of the more common dangers that may be found on a property. Contact your local extension agent or large animal veterinarian, or look online for a more complete list.

Here are just a few of the plants you need to look out for:

- Boxwood
- Horse nettle
- Larkspur
- Hemlock
- Nightshade
- Poppy
- Milkweed
- Lily of the valley
- Wild black cherry
- Sorghum
- Elderberry
- Black locust

There are also some plants that, because they are **photo-sensitizing plants** (containing properties that allow the plant to interact with sunlight), can cause your goat to become very susceptible to heat stroke and/or sunburn. The most commonly known plants would be Saint John's wort and buckwheat, although there are a number of others. It should be noted as well that Saint John's wort can also cause miscarriage in both animals and humans, so it is a plant (and an herb) that, should you decide to grow it in your garden, needs to be grown and kept safely.

will make short work of it, while enjoying themselves at the same time. Typically, the goat owner will set up a temporary fence around the area to be cleared and move the goats and the fence until the area is cleared or as clear as the goats can get it. Once the goats are done, both the goats and fence are removed, and the new project plan for the space can proceed. Of course there *will* be some human work to do as well, but you will be surprised by just how much work the goats can get done for you.

Hay Isn't Just for Horses

A staple in your goat's diet, especially if there is little browsing/pasture area available, is good-quality **hay**—up to four pounds per day per goat, depending on the size of the goat. Although there are a number of types of grasses that can be used for hay, there is really no "preferred" hay out there for goats, unless it is your goats' personal preference. As long as the hay has been properly harvested and stored and is mold and mildew free, any type of hay may be fed to your goats.

However, **alfalfa** (which is also a variety of hay) is usually fed to goats only in moderation. This is due to the fact that alfalfa can not only cause bloat, but is also very high in protein. The average home herd does not need all the additional protein that feeding straight alfalfa would provide. In fact, if your goat gets too much protein in his or her system (usually through extreme ingestion of protein) it can cause kidney damage. So if you give your goats alfalfa, please do so in moderation, perhaps in the form of a treat; a little in the morning with the regular feeding will go a long way.

One thing that you will notice, as it will probably frustrate the devil out of you, is that goats are notorious wasters of hay. It will seem that for every mouthful they actually eat, two or three will end up on the ground to be walked over and ignored. Sometimes the "spilled" hay may be salvaged and given back to them to eat. Many times, however, it has not only been walked on, but also urinated and defecated on as well. When that happens, the hay cannot be given back to the animals to eat and should just be left on the ground for bedding until cleanup.

There are ways to help protect against hay waste. Hay racks, which are dispensers that look like metal half-baskets, come either round or square and can be attached to the wall at the goats' level. Although hay racks will not stop the waste, they will allow you to keep the fresh hay up off the floor, at least until the goats get to it, and help eliminate some of the waste.

Mangers can be another option for hay. A manger is basically a food trough. Although this will keep the hay off the ground, the goats (especially the babies or pygmies) will tend to climb in and eat. While this may seem adorable, and you may get a few cute pictures out of this, it is not a good practice to encourage, as the goats standing in the manger may indiscriminately use it as a toilet—as they eat. This in turn will make the rest of the hay inedible.

Providing a hay rack or hay manger for your goats will help you to prevent your goats from wasting more than they eat.

While hay racks and mangers are usually for indoor feeding, you can also purchase or make hay feeders to use outside. Hay rings hold the big, round bales of hay that you see in many fields and barnyards. However, unless you have a lot of goats, you probably would not use these big rounds. Instead, you would use smaller bales. You can purchase smaller rings as well as covered hay-feeding stations, which are nothing more than a ring or manger with a roof. This is nice in that it will protect your hay from the elements, especially from the rain and snow. Conditions like persistent dampness is what leads hay to mold, and moldy hay doesn't just make your goats sick, but it can kill. Usually a goat will refuse moldy hay, but this is not a given. This is why it is so important to check hay for mold, especially hay that has been subjected to the elements, even if only for a few days.

Grains

Goats should also have a good **grain mix,** which can come as a mash or in crumble or pellet form. This will fill in any nutrient, vitamin, or mineral deficiencies that the goats could have in their basic diet. It is important that they keep this balance, as a lack of something could actually cause your goat to have adverse effects, much like a human body would when lacking in an important nutrient. You can get pellets that are specially milled for meat and dairy goats as well as for does, kids, and bucks. Your feed merchant will be a great help in making decisions based on your specific needs.

A treat that your goats will love is **sweet feed**. This is simply a pellet/green mix with molasses added. It is like candy to them, and I've never seen a goat refuse it.

In saying how important the grains are (no matter in what form), it is also very important to remember to *never* leave bags of any type of grains, feeds, or pellets where goats can get to them. Grains and feeds are one thing that goats will gorge themselves on, resulting in bloat (discussed in Chapter 11), which can be very serious, even resulting in death.

Other Nutrients

Finally, goats should be provided with **salt** mixed in with their minerals. This may be purchased in block or loose form. Some say the loose form is better than the block form, as they feel the block form does not allow the goats to get enough salts or minerals. Others feel the block form works just fine, so it will essentially be your call.

Just like with any animal, your goats need the right food. But too much of the right food can end up putting your goat into a dangerous situation. Just remember: as long as you have good feed-management practices, there should be few issues.

CHAPTER 5

BREEDING

If you are just keeping your goat as a pet, then most likely you will not be worrying about breeding. However, if you are depending on your goats to provide you with milk and/or meat, then you will need to consistently breed your goats.

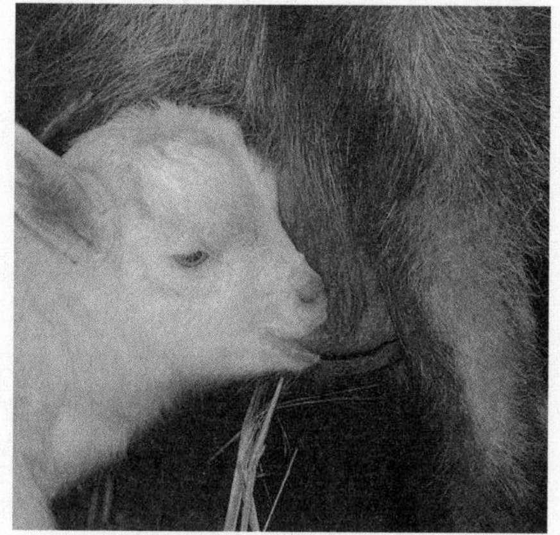

Your goats will produce enough milk to fulfill your needs *and* those of their kids.

Keeping Bucks

Whether or not you want to keep your own buck is primarily up to you. They can be a handful and are known for their **"buck" odor,** which is a turnoff for some goat owners.

If you are keeping a breed of goat that is rare or endangered, then most likely you will want to keep a buck on hand. You might also consider keeping a buck if there are no bucks in the area available to service your does. Additionally, if you have (or want) a **closed herd** (meaning you do not bring others animals into your herd), then you would keep a buck (or two). Don't be afraid to seek help in selecting your bucks, as you will need good stock to produce good offspring. A reputable breeder will help you choose the right buck for your breeding program.

But what if, for whatever reason, you cannot or do not want to keep a buck? There are basically two options available: either find someone with a good-quality, healthy buck who can service your does or use **artificial insemination** (AI). Should you choose either of these two options, it is important that you do your homework and ask questions. Visit the farms of potential sires, and don't be afraid to sit down and talk with the artificial-insemination companies as well.

Once you have your breeders in place, the next questions are when does the breeding take place, and how young is too young for goats to breed.

Breeding Season

Full-size dairy goats may breed as early as **eight to nine months old,** provided they have reached the proper weight, which is about 80 pounds. Some owners may want to wait a little bit longer. Just like dogs, cats, and all other mammals, does will come into heat, meaning that they are ready to be bred. Some goats, like the Boers, Pygmies, and Spanish (basically, any breeds originating in tropical climates), can breed at any time throughout the year. However, most of the dairy breeds, also known as the *Alpine* types (those originating in colder climates), will have more specific breeding seasons. This breeding season usually goes from August to January, with the does in heat every twenty-one days. The gestation periods of most goats are between 145 and 155 days, which is approximately five months.

Once you are ready to breed your does (or have already bred them), there are gestation tables and apps online to help you to quickly figure out approximate due dates for your does (you can find a few examples in the "Resources" section of this book). Keep in mind that if you are not going to keep the kids for yourself, regardless of whether they are milk or meat goats, you need to have a plan for the kids. It would help to put the word out that you will have kids available for sale. However, I would not advise guaranteed presales of kids, due to the fact that even the best-managed farm will have losses, and you don't want to be in a position of having to refund money that you have already spent.

As the does get close to their due dates, if you have the stalls available, you may want to begin bringing the girls in. There are a few reasons for this. First, if the does are kidding in late winter or early spring in an area with harsh weather, a thickly padded stall will help keep the newborn kids warmer than if they had been born outdoors. Also, the newborn kids will be quite vulnerable to attack. Remember that goats are prey animals, and that the littlest and weakest are usually taken first. Finally, it will help you to keep track of the doe and kid, making sure that both are healthy, are eating well, and are having no problems that need attention.

Now, when do you put the does and kids out with the rest of the herd? This is totally up to you, but when making your decision, look at weather conditions, how the does and kids are doing, and whether or not you're having predator problems. In the case of problems in any of these areas, does and kids should be locked up at night (as should the rest of the herd until the threat is over). When you finally do let the does and kids out with the herd, don't be afraid to bring a few back in if you see problems. Better to be safe than sorry.

Although many goats can and will kid on their own with no problems, every once in a while, problems will arise before, during, or after the birth. These problems can affect the doe, kid, or both (and, as luck would have it, these problems will usually be during the hours when your veterinarian is closed). Many owners like to try to be around during kidding time in case of problems, but this is not always possible. During kidding time, or at any time really, it doesn't hurt to keep an emergency first-aid kit available for use with the goats, as they have a knack for getting into trouble "after hours."

Quick Reference Guide: Goat-Medicine Supplies

The following is a list of items that are helpful to have in a first-aid kit for your goats. This list comes compliments of "The Goat Lady" at goatlady@goat-link.com.

Basics (Available at Drugstores):

Must-Haves:

- Thermometer (a digital one intended for humans is just fine)
- Box of baking soda (for rumen pH)
- Cayenne pepper (for stopping bleeding from injuries)
- Molasses or Karo syrup (for kid-saver formula)
- Peroxide (for cleansing wounds)
- Alcohol (for cleaning injection sites, needles, and the tops of injectable bottle stoppers)
- Milk of magnesia (for constipated goats or goats that have eaten something you need to get rid of)
- Pepto-Bismol (in small amounts for coating the stomach lining; too much will stop them up)

Nice-to-Haves:

- Antibiotic ointment, like Neosporin (for injuries)
- Baby aspirin (do not use ibuprofen, Tylenol, or Advil; only real, 100 percent aspirin, regular strength if you can't find baby)
- Gauze for wounds (if you can find cast supplies, get them; if not, gauze and plaster of Paris will work for broken legs)
- Benadryl (for allergic reactions to bee stings, bites, and hay-allergy breathing problems; "Children's" is fine, either as capsules or liquid)

- Coffee (for kid-saver formula)
- Surgical gloves (can be found in pharmacy departments; larger packages can be purchased at online supply houses)
- Yogurt with live bacteria in it (for rumen flora; to use in case you do not have probiotic paste)
- Olive or corn oil (for treatment of enema cases that do not respond to warm, soapy water)
- Tide Powder Detergent (for treating frothy bloat; add one teaspoon to one half-gallon of water, and give orally)
- Vaseline
- Diaper ointment or hemorrhoid cream (for topical skin abrasions; the hemorrhoid cream reduces topical swelling from a difficult kidding, and the diaper ointment is great for pizzle rot)
- Hair-color or hair-perm bottle (for administering enemas)

Injectable, Over-the-Counter Meds (Available at the Feed Store):

Must-Haves:

- Bottle of penicillin g procaine for injury and infection (dosage is one cc per 25 pounds, administered subcutaneously; always draw back on the plunger before injecting penicillin)
- Bottle of vitamin B complex (fortified if you can get it, because it has more thiamine in it; for stress and getting well, with the dosage is one cc per 25 pounds, administered subcutaneously)
- Bottle of Tylan 200 (not LA-200; for pneumonia and pink eye, with the generic being called tylosin 1 percent (not Tylan 50, either), and the dosage being one cc per 25 pounds, administered subcutaneously, or, for pink eye, being with the needle taken off and dripped into the eye—a couple of drops per eye)

- Tetanus antitoxin, equine origin (for injury, disbudding, castration, and puncture wounds, and as a preventative for tetanus; comes 1,500 units per vial and has approx 3.5 cces in it
- CD antitoxin (which is not the same as CDT; used for enterotoxemia symptoms in babies and adults, and can be a life saver to a potentially fatal situation, with the dosage being one cc per five pounds, administered subcutaneously)
- Ivomec Plus Dewormer (used at the rate of one cc per 40 pounds, administered subcutaneously)

Problems During Breeding

There are problems that you can run into with breeding. Some can be serious; others may be nothing more than the doe just not, for whatever reason, liking or wanting the buck. If you're using AI, maybe it hasn't taken, which can, sometimes, be an indication of a deeper problem, but other times, it is no different than a natural breeding that just hasn't taken. Some basic reasons a breeding may not take are:

Body condition: If the doe is too thin, it may not reproduce and may have problems with weaning (when kids stop nursing and go onto regular food). If the doe is too heavy, then toxemia in pregnancy may be a problem.

Missing the heat cycle: This is usually not a problem with the goat breeds that have no set breeding season, unless you are trying to specifically time a breeding. But for those breeds that do have a set season, missing the signs will also mean missing the breeding opportunities.

As does will go into heat every eighteen to twenty-one days when in season, there are signs to watch that will clue you in as to when your girls are in heat.

Some of the signs are:

- Continual tail wagging
- Lots of vocalization
- Slightly swollen vulva
- Discharge around the tail leaving the area wet
- Pacing if she knows a buck is near (that "buck smell")

If you need to stimulate your doe, use a **buck rag.** This is simply a cloth rubdown of the buck's scent gland, used to detect or induce heat in a doe. Store the rag in a jar when not in use.

These are just a few signs that you should be looking for. You can read more in-depth information about your does and their heat cycles online, through extension office information, and in the "Resources" section of this book.

Sometimes it is the buck that has the problem. One of the problems you may experience with your bucks is his sperm count. However, this can be corrected the more the buck tries to breed. A low sperm count in a buck is usually a result of the buck not having been used in a while. Conception rate will be low while the buck makes his comeback and his sperm count rises.

Another problem could be that the buck just doesn't want to breed. This can be remedied by simply not keeping your buck(s) and does together.

Of course, these are not the only breeding problems you may face; these are just some of the basics. Again, experienced breeders, extension offices, and veterinarians are excellent sources of more in-depth information on breeding problems.

Problems During Kidding

Even with successful breeding, problems can still arise before, during, and after the birth of the kid. (Only problems, not solutions, will be listed here. Before your first kidding, it is strongly advised that you read up in depth about the possible problems the doe may face.)

- Kid has large head and shoulders
- The kid's head is out, but one or more of the legs are not
- The kid's hips are stuck
- Breech birth
- Afterbirth does not pass
- Infection of the uterus

These are only a few examples of birthing and afterbirth problems and complications. If your goat has been having problems for thirty minutes, and nothing you do is helping, the veterinarian needs to be called. Having said this, when your doe does begin to kid, don't be afraid, and don't assume the worst is inevitable. In most scenarios, the doe and kid will do just fine. But you do need to be prepared.

Problems During Pregnancy

Another problem you may face is a doe aborting her kid. A few reasons for this are:

Brucella: Considered to be nonexistent in the United States, brucella can happen through consumption of contaminated food, including through pasture and/or water.

Toxoplasmosis: Caused by cats leaving feces in the hay or feces getting into water or feed.

Listeria: Caused by bacteria found in contaminated water, spoiled feed, or even from the soil, where it can survive for quite some time.

Although there are many other reasons for an abortion to occur, these are some of the most common. Solutions to these problems have not been included here, as any of these conditions will require veterinary attention.

If your goat *does* abort, you need to look into the reason why. If you are suspicious of a serious problem with the goat or herd, contact your veterinarian immediately. They may ask you to refrigerate the fetus and placenta; if so, handle each with care, using disposable surgical gloves.

One more reason for abortion is poisonous plants. Check the pasture for something the animal may have gotten into. Should you discover that a poisonous plant is, indeed, the problem, take a plant sample and call your veterinarian in case further action needs to be taken with your doe.

Although up to a 5 percent abortion rate is not uncommon in a herd, it is still a serious situation. A higher rate can mean devastation to you and the herd, and is most likely the result of an infectious problem. Should you, as a first-time owner, experience your doe or does aborting, call your veterinarian; at the very least, contact a very experienced breeder to guide you through the possible problems, as well as their solutions.

Breeding Cycles and Older Goats

One final problem in kidding is one that is totally in your control: acting when a goat is too old to breed any longer. A doe, no matter how healthy she is, should be retired from breeding at around 10 years of age. This is simply due to the fact that the older the doe gets, the harder it can be for her to kid. Some have had does up to age fourteen still kid, but do notice that it gets more difficult for them. And despite these older does kidding, the average **"retirement age"** seems to be around the 10 year mark. Also, when an owner sees that a good doe is nearing or at her retirement age, he/she may keep his/her final kids (especially any females) to keep his/her line going (new jobs for retired goats will be discussed in Chapter 12).

Your Goat's Needs During Pregnancy

Your does will have special needs during their pregnancy and after they kid. Early on, during the first three months, does should be fed to maintain their body condition (if it is good); otherwise, you should take this time to improve their body condition. A little extra grain will help an adult who needs some weight on it, but do not overfeed, as that can cause such problems as ketosis. Extra hay can also be offered.

In the last two months, your doe's nutritional requirements will dramatically increase, due to fetus growth. At this point, grains or milking rations may be increased, but slowly and in small amounts until the does are eating up to a pound and a half per day (depending on breed/size) by kidding. Dairy does should be fed at least half that number, up to two thirds their other milking rations. However, if does are given too much grain, it could result in the fetuses growing too big, causing problems at birth, so monitor accordingly. Continue to feed them hay; however, alfalfa can also be fed at this time and can even gradually replace the hay.

Throughout your doe's pregnancy, continue to monitor her. If something looks wrong with her body condition, adjust her diet as necessary. During this entire time, her water intake will also greatly increase. Be sure she always has plenty of clean drinking water available.

After the doe has kidded, her grain rations should slowly be increased, reaching three pounds of grain per day by the fourth week after her kidding. When her lactation peaks, feed her according to her milk production (in the case of dairy): a half pound of grain per pound of milk (over three pounds). Also, feed her good forage as well; a goat should never have more than four pounds of grain per day (this applies mostly to dairy breeds).

Breeding and kidding time are very busy times in a goat owner's year. And although things can go wrong, when it goes right, the benefits are considerable. The kids are some of the cutest little things that you will ever see, while a good buck and doe are worth their weight in gold. The first few births may seem scary to a new goat owner, but you will become more comfortable with the process with each additional birth. And while the first complication that occurs will be another scary time, don't panic; if you freeze, you could lose both doe and kid. If you feel you can't handle the situation, call your veterinarian. But just as with the births, with each "emergency" that passes, you'll find that you handle it better and better. Bottom line: don't let breeding time overwhelm you. Enjoy it and the resulting kids.

CHAPTER 6

KIDS

As we discussed in Chapter 5, after the breeding comes the kidding. When all is said and done, you will end up with a barnyard full of beautiful little babies ready to take their place in your (or someone else's) herd. Once the kids are born, they will start out nursing from the doe. While meat breeds will nurse until they are ready to leave, dairy kids will nurse for a much shorter time.

If you are raising meat goats, unless the doe cannot nurse for some reason, then the kids can remain with the doe until weaning. However, if you are raising dairy goats, you will want the milk for your own purposes, so the kids will not be able to get the milk they need. As a result, kids of dairy goats will need to be bottle fed. (However, some goat owners have found a happy medium in milking for their use only once per day and letting the kid stay with the doe. This usually works

Newborn goats will nurse for as long as they are able; it will be up to you to decide when to start weaning.

only for owners wanting milk just for family use and having a doe with a good milk supply.) Whether for meat or dairy, it is important that the new kids nurse from their mothers for the first few days in order to get the necessary colostrum.

Colostrum is the first milk the mother produces. It is very rich in nutrients and protective antibodies, and it gives the kids' digestive system a jump start. It is best that the kids get the colostrum they need directly from their mothers; however, there are times when this may not be possible, such as when the doe is ill and has contaminated her own milk. When this happens, there are some alternatives. Look for a local provider of frozen colostrum. If you're not aware of any, ask your veterinarian if he or she knows of a provider. If this is not an option, then you may use a commercial colostrum replacement. It should be available at most farm/livestock supply stores. After a few days, whether using your doe's own colostrum or replacement, transition the kid to a bottle-fed, commercial milk replacement, which should also be available at farm/livestock supply stores.

Bottle feeding kids is not difficult, but it is time consuming. Nursing is pretty much instinctive for kids, so you should have no problem getting them to take to the bottle. If they don't want the nipple at first, gently open their mouth, put the nipple in, and squirt a little milk into their mouths. They will quickly get adjusted and start sucking.

It is important that your kids receive the colostrum they need to stay healthy; if they can't get it from their mothers, they will need you to provide it.

For the first three weeks, the kids should be fed at least four times per day. Starting at week three, you should also begin to slowly introduce solids into their diet as well. At week four, bottle feeding can be cut to three times a day, with free choice of a 14 to 16 percent protein feed. Starting the kids on solids will begin the stimulation of rumen (first stomach) development.

One thing that you need to watch for in kids being fed milk replacement is bloat; especially if the replacement contains soy (bloat will be covered further in Chapter 11). For this reason (and others), it is a good idea to make all food changes gradually.

Bottle-fed goats may be weaned at eight weeks, as long as they are eating their hay and grain properly. Kids that are feeding directly from their mother (meat-breed kids) may be weaned at any time from eight weeks on, after which you should monitor their body condition and food intake. Some goat owners will wait a little longer before they wean the kids of meat goats.

When you wean a kid from its mother, you may hear a lot of bleating and crying, especially from the kid, but at times from the mother as well. Weaning can be a stressful time for the kids, so you will need to watch for **weaning shock.** With weaning shock, kids (with males being more susceptible) can slow or even stop their growth. They may also lose weight. To minimize the problem, just be sure that the kid is the proper weight for the particular breed before weaning; the healthier the kid, the less the risk of and impact from the shock.

Although bottle feeding does take some time, it can be a necessity under various circumstances. But beyond that, it does allow you to spend time with and learn about the kids and bond with them if they are to stay with your herd. If done properly, you will have just as healthy a kid as if it nursed from the doe.

Wethers

Wethers are simply castrated males. Unless you know for sure that the male kids will be used for meat (at which point they can be castrated in a week or two) or breeding purposes, they should be castrated at about eight to twelve weeks. Should you choose to keep a male kid for a pet or as a companion to another animal, if you don't castrate, then you can end up with a goat that could be difficult to handle and could smell. A castrated male, on the other hand, will be a sweet-tempered pet. This is not to say that it is required; I did have a very sweet-tempered breeding buck with little smell named Elvis, who I kept until his natural death. However, this buck is the sort of exception that proves the rule.

Castration is not difficult, and with a little something to help the kid out with the momentary pain, you'll both walk away in good shape. However, castration techniques will not be described here. Rather, for your first castration experience, it is highly recommended to have either an experienced goat owner work directly with you or your veterinarian visit and teach you.

Disbudding

Many dairies will disbud their kids; this is to prevent their horns from growing. A kid may be disbudded by using a specially created hot iron or caustic paste, usually within a few weeks of birth. However, both methods need to be done properly, or other problems may occur, including burns from improperly used paste and disfigured horns from improper or failed disbudding. Again, this process will not be covered in the book. To properly learn about the disbudding of kids, contact a very experienced breeder or your veterinarian and have them teach you on-site.

This is just a small sampling of the kidding experience. You will gain more confidence and experience with each kidding year. And if you have even an inkling that something is wrong, don't be afraid to call the veterinarian. But after your first experience, I guarantee you'll become hooked on the little kids. Enjoy them while you have them.

CHAPTER 7

MILKING YOUR GOATS

When people keep goats on the farm or homestead, most are keeping them for **milking.** Although some may keep a couple of dairy goats for pets, for work, or for packing (discussed further in Chapter 12), the majority of farms, homesteads, and households keep dairy goats for their milk.

Keeping milk goats will require you to set aside space and equipment that you wouldn't need with meat goats. It also means that you must keep a schedule, as your goats must be milked at the same times each day. And in order to be able to milk, the does do have to kid, usually every year. I say *usually,* as some goats (and some goat breeds) can continue to milk without kidding every year. In fact, I knew of a goat that was still milking well over three years after its last kid. While this is not the norm (and yearly kidding is), it is something that *can* happen, but that should not be an expectation.

When you have milk goats, especially several of them, you will need a **milking parlor.** This is simply a separate space from where the goat stays that holds the milking equipment and serves as a clean space to milk your goats. This room can be part of your barn; it does not have to be a separate building. If you do not have a separate room in your barn, you can select an area and wall it off. If you are only milking a few goats, a single-door room would be plenty. In this way, you can create your milking parlor without having to add on. Just make sure you allow plenty of room for the equipment.

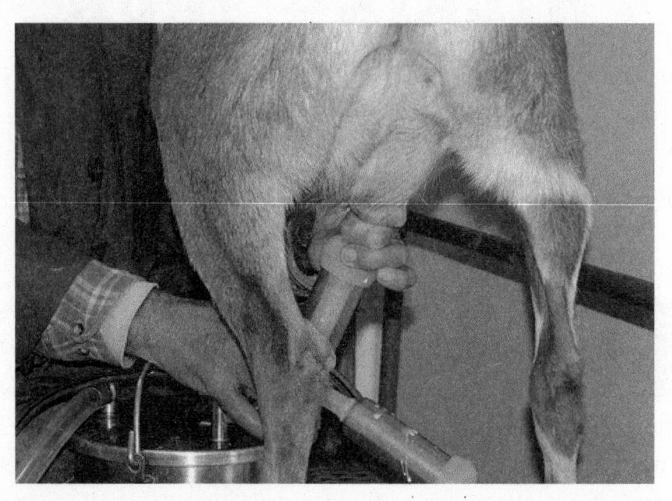

Milking your goats can seem difficult or confusing at the outset, but the proper equipment will make it easier and more manageable.

When to Milk Your Does

A doe's milk production will change as time goes on. Its prime milking time will be **from kidding time to six months after kidding,** when, depending on the goat, it will produce up to a gallon of milk per day. At six to eight months following kidding, does can produce approximately two to three quarts of milk per day, after which they may go down to approximately one quart of milk per day. (A Nigerian Dwarf Goat, for comparison, will produce approximately two quarts per day at peak.)

Once the milking is done, the milk must be quickly chilled to below 40°F. This must be done within the hour. For small amounts (from just a few family goats), an ice bath should work fine. Putting the fresh warm milk into the refrigerator will not cool the milk down fast enough. There are commercial coolers available, but unless you are cooling six or more gallons a day, this is not a necessity. However, it will make the cooling process go much easier for the larger amounts.

Equipment

When you milk goats, whether one or one hundred, you will need equipment. The following is a basic list of what any milking parlor needs:

- Milking stand
- Strainer/filters
- Milk pail
- Sanitizer
- Containers (glass)
- Teat dip and cup
- Udder cream
- Paper towel
- Milking machine (optional)
- Cream separator (optional)
- Pasteurizer (optional)

Now we'll take a brief look at each piece of equipment.

Milking stand: This is a platform that the doe stands on to be milked. It will be up off the ground, making it easier to milk her. It has a keyhole stanchion that holds her head and a feed dish on the front so she can eat as you milk. Milking stands may be metal or wood, and can be either purchased commercially or built on the farm (links for plans are included in the "Resources" section of this book.)

Once a doe is accustomed to the routine, it will usually climb right onto the stand and put its head into the stanchion on its own (of course with feed in place) and quietly nibble as you milk.

Strainer and filters: Used to strain/filter the fresh milk. Stainless steel is best to use, as it is easiest to clean and sanitize. However, some have found that canning funnels with reusable coffee filters (that can be thoroughly cleaned and sanitized) or disposable paper filters work as well. Some still use cheesecloth for straining.

Milk pail: Usually a small, seamless pail used to hold the milk. If milking by hand, you will milk directly into the milk pail. They are easy to clean and sanitize.

Sanitizer: Used on teats and udders before and after milking.

Containers: These will be used to store your milk. Most keepers prefer glass over plastic, as the plastic can affect the milk's flavor, and some say that the milk will not last as long. For glass containers, even mason jars will work.

Teat dip and cup: Used on the goats' teats after milking; with the teat cup, the teats are dipped into the liquid that has been poured into the cup to prevent infection and heal the teat. This helps prevent against mastitis (see Chapter 11).

Udder cream: Used on udders after milking. Prevents chapping and promotes healing of some skin problems.

Paper towel: Used to dry udders and teats after sanitizing.

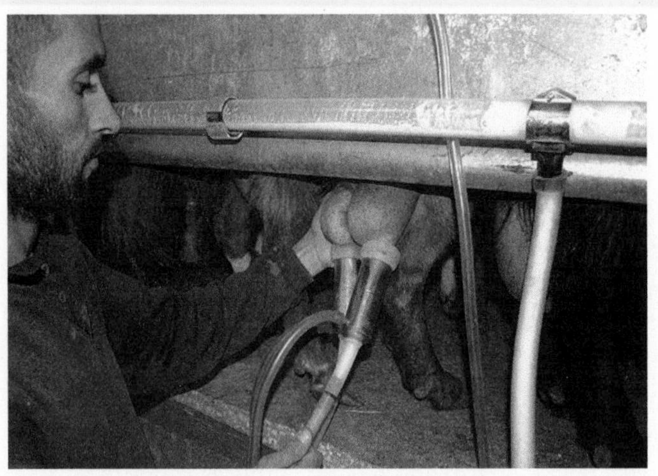

Commercial goat milking, which focuses on production, uses many of the same techniques the backyard farmer will employ.

The following are options for milk-house equipment:

Even an antique cream separator makes for a valuable investment.

Milking machine: Unless you have a barn filled with goats or, for some reason, cannot milk by hand, a milking machine will be an expensive purchase that really is not necessary. Basically, a milking machine has cups that attach to the goat's teats and, through a vacuum that is created by the machine, take the milk from the teats, flowing through attached hoses into an attached container (typically stainless steel).

Cream separator: If you want to make butter or want the fresh cream for anything else, a cream separator will make the process of separating the cream from the milk much easier. Although this can also be done by hand, if you can afford a new machine or find a good used one, it will be a worthwhile investment.

Pasteurizer: If all you will be drinking is raw milk, you will most likely not need a pasteurizer. But if you are concerned about bacteria, then you'll want to gently pasteurize your milk. We will discuss pasteurized milk further in Chapter 8.

All of the equipment mentioned is quite easy to find at farm stores and online stores that cater to goats and their owners. Although some of this equipment can be safely purchased used, you should go over it thoroughly before making your final decision.

CHAPTER 8

GOAT MILK AND ITS USES

Milk is the "white gold" that all who raise dairy or milk goats is after. Goat's milk may be used just as cow's milk, and is even used by those individuals whose systems cannot tolerate cow's milk. Because a doe will not begin to lactate unless it first has a kid (as discussed before), the doe must be bred before it can be milked. The kids will nurse just long enough to get the all-important colostrum, and then will be bottle fed until weaning. However, some goat owners claim that they can still allow the kid to nurse on the doe once a day and have enough milk for their own use. This would most likely depend on the doe first and how much milk it gives per day, along with how much milk the goat owner needs for his or her own use. Others will not start milking until they have weaned the kids, at which point they will immediately begin to milk the doe so it doesn't dry out.

Goat cheese is one of the most popular uses for goat milk.

Raw Versus Pasteurized

Once the milking is done, the milk must be quickly chilled to below 40°F. This must be done within the hour. For small amounts (from just a few family goats), an ice bath should work fine. Putting the fresh warm milk into the refrigerator will not cool the milk down fast enough. There are **commercial coolers** available, but unless you are cooling six or more gallons a day, this is not a necessity. However, it will make the cooling process go much easier for the larger amounts. Once the milk is cooled, you will now need to decide whether you want to drink/use your goat's milk raw or pasteurize it. This is essentially a personal choice.

Pasteurization involves the heating of the milk to deter bacteria. It can also help to eliminate the "goat" flavor that the milk can acquire if stored raw. However, pasteurization can also eliminate the good things in the milk. Because of this, some who

Goat cheese, when allowed to age, is a delicious addition to dairy recipes."

pasteurize feel that thirty seconds at 160°F is better than thirty-five minutes at 145°F. It is also said that the milk has a better flavor when cooled quickly after pasteurization.

All the same, raw milk is becoming more and more popular with people. Raw milk is just that: milk that has not been pasteurized, retaining all the natural nutrients that make the milk healthy, as well as having a better taste. Those against raw milk claim that it is dangerous because any bad bacteria that the milk may contain are not killed. However, if your equipment is clean and sanitary, your goats are well kept and healthy, and you follow the cool-down and storage processes properly, this should be a nonissue.

Whichever method you choose, just make sure that the entire process from start to finish is followed correctly, and you should run into few problems.

Draining the whey from your goat milk allows you to make use of it for other projects.

Storing Your Milk

As discussed before, milk is best stored in glass. You can use bottles or even mason jars. It goes without saying that the milk and cream (if separated) need refrigeration; that being said, if you are overloaded with milk, you may also freeze or can it. Should you decide to freeze your milk, remember to leave room in your container for expansion. Do not fill containers to the top. And, although the use of plastic is not recommended for storing milk, if you are not comfortable freezing in glass, then don't.

It should be mentioned that freezing milk actually keeps a fresher taste than canning, but if you need freezer space for other foods, you may find yourself needing to can your excess. When canning your milk, it should be fresh and not stored in the refrigerator for a few days beforehand; otherwise, there is a risk of curdling when using a pressure canner.

Goat milk whey is used in a variety of delicious recipes.

Uses for Milk and Cream

Goat's milk also has many of the same uses as cow's or sheep's milk. You can drink it, cook with it, or make cheese with it. You can also make ice cream, yogurt, and kefir. See the "Recipes" section at the end of this book for some examples. However, some recipes will work best with raw milk, some may not be able to use pasteurized milk, and some may work with only fresh milk, not frozen or canned milk. If you separate the cream, you may also make butter and/or whipped cream.

On the nonfood side, you can also make soaps and lotions from goat's milk.

Although milking your goats means a commitment from you, they will reward you with some of the best milk that you have ever tasted. Just remember that cleanliness for you and your goats, as well as your equipment, is paramount to a top-quality product.

Goat cheese comes in numerous varieties, depending on how it is prepared and aged.

CHAPTER 9

MEAT

Although any goat may be used for meat, some people raise goats specifically for meat and have preferred breeds for doing so. As discussed in Chapter 2, meat goats will be of a larger build and will be more muscular and meaty than the dairy goat will be.

Goat Meat

Goat meat is quite good and is the most consumed meat worldwide, eaten by nearly 70 percent of the population. It is lower in fat and cholesterol than beef, yet is higher in protein. The cuts are similar to lamb, yet goat meat is lower in calories than beef, chicken, pork, or lamb. And if you are raising them, you can put more meat goats on an acre of land that you can beef cattle.

This two-week old kid is an example of a cabrito goat.

According to the United States Department of Agriculture, the meat-goat industry is one of the fastest growing in the

United States. As of 2008, there were more than 3 million meat goats in the United States, and that number continues to grow. One of the reasons for this is the rising immigrant population from cultures in which the primary meat is goat, which, in turn, is resulting in more Americans discovering the healthier benefits of goat meat. It is also a versatile meat, good for almost any use that the chef or home cook can put it to.

You will notice that goat meat is marketed in different ways. **Chevon** is meat from goats six to twelve months of age, while **cabrito** (meaning "kid") comes from the young goat, usually anywhere from one to three months old and usually milk fed.

And although in the United States we usually think of sheep meat when we hear the word mutton, it is also used in reference to goat meat (chevon).

Some of the various cuts of goat meat include:

- Shank (lower leg)
- Ground meat (burger)
- Stew
- Chuck roast
- Rib rack
- Tenderloin
- Chops

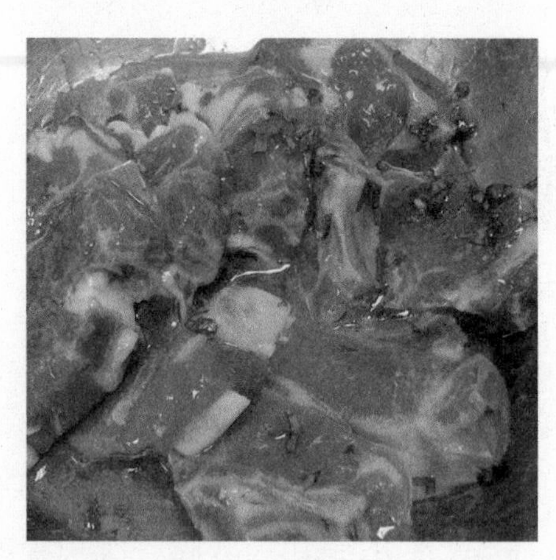

Goat meat is considered a delicious staple in the majority of world cuisines.

Although some grocery stores may carry goat meat, it is more common (and preferable) to purchase it directly from the farm or a butcher.

Slaughtering Your Goats

When raising goats for meat, some owners who raise for their own meat prefer **on-farm slaughter,** even to the point of doing it themselves. This is less stressful on the goats as well as the owner, as they feel (and are usually correct) that they have full control and can make sure that the goat is slaughtered quickly and humanely. Should you choose on-farm slaughter, make sure that you learn the right way to do it. Nothing is more tragic than slaughtering an animal and not doing it properly. It can turn a well-meaning action into a cruel end of life for the animal.

If you don't already know someone, then find someone who is proficient at home slaughter. Talk to them; ask them questions to make sure that they will teach you according to your standards. Ask if you can come and observe when they do one of their own. If their methods meet your expectations, then hire them. If they do things that you don't like, or if you feel that they are not treating the animal respectfully, thank your host and move on to find someone who fits your needs.

What if you want on-farm slaughter to ensure that you control what goes on, but do not want to do the deed yourself? There are people who will come to your farm to perform slaughter and butchering. Many times, other livestock owners may know of someone who does this. If not, your local extension office may have information available on those who do on-farm slaughter. Just be sure that they know how to slaughter goats.

If you sell large numbers of meat goats, then sending the animals to a local **slaughterhouse** may be the way that you need to go. If you are not totally comfortable with sending your animals off of your farm for slaughter, don't be afraid to actually interview the people at the slaughterhouse. Talk to some of their other clientele. If you want to be sure that the animals are being handled humanely and respectfully, ask if you can see the process. Again, if you like what you find out, hire them. If not, thank them and move on. On a side note, many times small, family-run slaughter and processing operations will offer what you will be looking for more reliably than larger-scale outfits.

You will probably find that the first time that you have your goats slaughtered, whether you do it yourself or send them out, will be difficult. Although saying it gets easier with time is not quite the right thing to say, you do get used to the fact that it is part of raising meat goats. And if you know your animals are being handled properly, humanely, and respectfully, it will make slaughter time a bit less stressful for all concerned.

A quick note: all states have their own rules and regulations on selling meat and dairy products. Should this be your goal, you need to check with your state's regulations to find out what you will need to do.

CHAPTER 10

CULLING

Culling describes the act of removing an animal from its herd for a number of possible purposes. Although many think that culling involves killing those not wanted in the herd, this could not be further from the truth. In fact, **culling doesn't necessarily require slaughtering the animals.** Culling can, however, involve putting down an animal that is sick or hurt and that has no hope of recovery or sending an animal to slaughter if it is otherwise healthy, but is not right for the herd. In many cases, though, including with goats, healthy animals that are culled may be sold for pets and as pack animals, among other uses. Culling also helps you keep your herd manageable. If the thought of the culling process bothers you, just remember that culling does not primarily involve putting down sick animals. This is included, of course, as we have discussed, but it is not the primary function of culling. The primary function of culling is removing, in a variety of ways, unwanted animals, sick or healthy, from the herd. Culling being synonymous with killing is a huge misconception that many have and that prevents some herders from doing a necessary clean-out of their herd.

Culling is an ongoing process if you are keeping a herd of goats. In the beginning, it may be tough, and you may second-guess your choices. That's okay. Before you permanently eliminate the proposed culls from the herd, get a second opinion. Alternatively, you can have someone work with you the first time that you cull.

You will find that, as you learn more about your goats and what is working for you and them, the culling process becomes easier.

When Culling Should Be Considered

When culling dairy and/or meat goats, some things that you will need to look for will be the same, with slight variations depending on goat purpose. Let's look at a few situations that would call for culling from the herd:

No/low production: While this applies only to dairy goats, you will need to be sure that your goats produce enough milk for your needs, whether you are running a dairy or milking only for yourself. If the goat didn't produce enough milk for your needs, it would need to be culled, at least from your herd. However, it may work fine for someone looking for a single goat for his/her own needs.

Growth in size: This is mostly a concern for meat goats; meat-goat breeders depend on their animals to be fast growing and muscular to meet production needs, especially if the animals are to be part of a breeding program.

The following are examples of situations that would merit culling for both meat and dairy:

Reproduction problems: If you continually have problems breeding a doe or have a buck that will not perform or does not seem to be producing offspring, they may need to be culled.

Temperament and personality: Goats that have a bad temperament or an aggressive personality or otherwise cannot be handled safely can be a problem for you, especially if you have children who will be handling them.

Continual kidding problems: If the doe continually has problems when it kids, it may be a candidate for culling.

Chronic health issues: Goats who continually have the same health issues over and over present a problem. If it is not serious and is not something that can be passed to offspring or the herd, and you feel it is manageable, that's fine. However, if the problem could be passed on to others in the herd or offspring, or if the medical bills stacked up, then you should seriously look at culling. Another cause of a chronic health condition could be a weak immune system, in which case the goat would be susceptible to illness, with does falling ill regularly; then culling may need to be done.

Worn teeth: If the goat's teeth are too worn, the animal will not be able to eat and will starve (especially if foraging).

Udder and teats: Damaged or saggy udders would be looked at for culling. Saggy udders that drag can be ripe for infection over and over again.

Injury: Sometimes a goat can bounce back from an injury, and sometimes not. A bad injury that a goat cannot recover from would merit culling.

These are only a few examples of what you may consider when you need to begin culling your goat herd. You may, and most likely will, have other things that you will need to look for when making your final selections; all that will come with experience.

Options When Culling

But what about that one special goat that is too old to milk, isn't a meat goat, and is too old to produce kids any longer? It is no longer "useful," but you can't bear to part with the animal. There is no problem or shame in keeping an older animal. Just be sure that you will be able to accommodate its needs and recognize that it may mean not being able to add a new goat into your herd for a while. If you're fine with that, then, by all means, keep the goat. If you find you can't keep it, but do not want to send the animal to the sale barn, either, find someone who wants a pet goat or maybe has another goat or even a horse that needs a companion. As has been said, culling does not mean you have to slaughter.

CHAPTER 11

DISEASES AND ILLNESSES

Although goats can be quite rugged, tough little animals, like any other animal, they can get sick. Sometimes it'll give hints or a helpful symptom will tell you that something is wrong. Other times, your goat could be in a very bad way before you even realized that it was ill. For this reason, it is important that you watch your goats and get to know their routines and personalities, as picking up on an off day could save your goat's life and you a lot of grief, aggravation, and even loss.

Some Common Diseases

The following are a few examples of what can affect and infect your goats.

Mastitis

Caused by bacteria, **mastitis** is basically an infected udder. Although meat goats may contract it as well, mastitis is usually found in heavy milkers. Symptoms of mastitis include a swollen, hot, and hard udder, milk that is stringy and specked with blood, and a decrease in overall milk production (if there is any milk at all).

Treatments include removing the kid (which you will need to bottle-feed), milk out the udder, and infuse the affected teat with intramammary medications or antibiotic injections and teat infusions. Hot packs and massaging the udder with peppermint oil can help increase circulation. You may need to treat for fever as well. Mastitis may cause permanent damage to the udder. Prevention involves keeping the udder and teats clean and the does' area clean. Mastitis can be chronic.

Ketosis

Ketosis is a buildup of excess ketones (made when the body breaks down fat for energy) in the blood. It will occur right before or not long after the doe kids. It is caused by the doe not getting the correct nutrients during its pregnancy. As a result, the doe's body will use its own protein reserves to feed the kid. Its own body then starves by producing more milk than it is actually able to handle.

Some signs that your doe may be struck with ketosis include:

- Teeth grinding
- Breath and urine have a sweet, fruity odor
- Dull eyes
- Doe separates itself from the herd
- Staggering
- Backs off of or completely stops eating

Treatment for ketosis is propylene glycol, molasses, or Karo syrup. Prevention is proper nutrition for the pregnant doe. Ketosis can be fatal.

Respiratory Problems

One of the most common respiratory problems in goats is **pneumonia,** which is associated with the lower respiratory tract of the goat, and can be due to virus, parasite, or bacteria. Kids are most often affected by the viral type.

The upper respiratory tract can have **sinusitis.** Causes may be nasal tumor and foreign objects in the goat's nasal passages, among other things.

The pharynx/larynx area can be affected by an abscess. These injuries are usually caused by improper administration of oral medications. Rather than the medications themselves causing harm, the equipment used to give the medications can harm the goat, as it usually means having to put something down the goat's throat to administer. When equipment is not used properly or is used too roughly or aggressively with the animal, it can result in injury. These injuries could end up interfering with the animal's ability to swallow.

Any upper respiratory problems can affect the breathing of the goat, and you need to contact a vet or experienced owner. New owners should not handle these problems until they are more experienced.

Mange

Mange is a highly contagious parasitic-mite infection. It is not deadly. Symptoms of mange include hair loss and rubbing and scratching on posts, feeders, etc. The scratching and rubbing could, in turn, result in open wounds. The continual scratching can also affect weight gain.

There are three types of mange:

Sarcoptic: This type is caused by the *Sarcoptes scabiei var caprae* mite. It lives under the skin in the head/neck area, where they burrow to lay eggs.

Psoroptes: Caused by the *Psoroptes cuniculi* mite. This mite usually spends its time in the ear region, and is also a burrower. This is called "ear mange," too.

Chorioptic: Caused by the *Chorioptes bovis* mite. This type affects the legs and feet, where crusted areas and lesions may form where the mites are active.

Most cases of mange spread through direct contact, and overcrowding will be a factor in rapid spreading. It should also be noted that this is species specific, meaning mange mites of goats cannot affect other animals or humans.

Ringworm

Contrary to the name, **ringworm** is a fungal disease, not a worm. A common skin disorder in goats, ringworm can be identified as a round patch with a hairless ring.

Treatment for ringworm is a thorough washing of the site with a topical skin disinfectant. Dry the site and apply Clotrimazole 1 percent cream, repeating daily until the ringworm is gone. Always use disposable surgical gloves when treating ringworm. Treatment may take weeks.

Ringworm is contagious, even to humans.

Poisoning

Common causes of **poisoning** are two things that have already been discussed: too much grain and toxic plants (Chapter 4). Good management is the best prevention.

Sore Mouth

Caused by the pox virus, **sore mouth** is viral, but needs some sort of break in the skin to actually get into the body. Symptoms include scabs/blisters on the lips, udders, teats, and/or nose.

Sore mouth can result in susceptibility to other diseases, starvation (as it may be too uncomfortable for the goat to eat), and condition loss. Sore mouth is transmitted through direct contact with other goats or anything that may have come in contact with the virus.

There is an injection available for sore mouth; however, it can also be treated with an iodine/glycerin treatment. Alternatively, it can be left alone and allowed to clear out on its own; keep in mind that you need to make sure that the goats are not starving due to a discomfort in eating.

White Muscle Disease

White muscle disease is a degenerative muscle disease, usually found in newborns, and caused by a selenium and/or vitamin E deficiency. It affects the skeletal and heart muscles, and, in older goats, may cause low conception rates, low milk production, and fetus absorption, among other concerns.

Treatment is an injection of vitamin E and selenium.

Pinkeye

Also known as *keratoconjunctivitis,* **pinkeye** is an inflammation of the inside of the eyelid. Caused by *Mycoplasma conjunctivae* and *Chlamydia trachomatis,* pinkeye is highly contagious.

Pinkeye can occur through the introduction of a new goat, relocation, transport, or even stress from weather conditions. Symptoms include red, swollen, and watery eyes, as well as squinting and/or cloudiness in the white of the eye. There can also be pain in and pus drainage from the eyes, along with crusting. If left untreated, pinkeye can cause permanent blindness, and temporary blindness may occur as well.

Treatment includes isolating the infected goat or goats from the rest of the herd. Flush the eyes of the infected animals using a sterile saline solution, and then apply an antibiotic ointment. Wear disposable surgical gloves when treating the animals, changing gloves with each animal.

Johne's Disease

Caused by the bacterium *Mycobacterium avium,* **Johne's disease** is a fatal gastrointestinal disease in which the intestinal wall inflames and thickens, resulting in the intestine no longer functioning properly. It usually happens early in the first year of a goat's life, but symptoms may not show for years. It can spread.

Symptoms include weight loss (even when eating well) and/or diarrhea. Herds can be infected when an infected goat, which looks and acts quite healthy, is brought into the fold.

The best control against Johne's disease is to avoid introducing it into the herd in the first place. Either purchase Johne's-free animals or maintain a closed herd (meaning that no new animals are introduced from the outside and that any new stock comes from your own herd).

Coccidiosis

Coccidiosis is due to intestinal parasites, usually affecting young kids with immature immune systems and recently kidded does due to body stress. Along with stress, it may be caused by overcrowding, wet pens, and/or dirty water.

Symptoms include, but are not limited to, diarrhea, dehydration, fever, and weight loss (which, should the goat survive, could become chronic).

Although coccidiosis is species specific, it is very contagious. Treatment is the use of coccidiostats, which will inhibit development. It's important to note that deworming will not help.

This is just a sampling of what can affect your goats. Some are treatable; some are not. In some cases, danger can be totally avoided simply through the good management of your goats.

However, as a new goat owner, until you can begin to recognize problems and familiarize yourself with treatments, it is best to contact either an experienced goat owner to help you with health problems in the herd or your veterinarian.

CHAPTER 12

PREDATORS AND OTHER DANGERS

As has been mentioned, goats—especially the kids and elderly, sick, injured, and miniature goats—are **prey animals.** This means that your goats will be looked at as food for other animals.

Types of Predators

Depending on where you live, the types of predators you'll need to prepare for will vary. Predator types range from mammals to reptiles, and even some birds will prey upon goats. Examples of these predators include coyotes, wolves, bears, alligators, constrictors, eagles, vultures, dogs (both domestic and domestic-feral—those dogs that were formerly domesticated, but have turned wild and can run in packs), foxes, wild cats, and wild hogs.

While all predators can potentially do great damage and kill, **wild animals** will usually kill only for food and will tend to take only what they need. However, **domestic-feral dogs** (and some domestic dogs) *will* slaughter, making them one of the more significant dangers to a herd. A pack of domestic-ferals will kill multiple animals, more often than not just to kill. In fact, in farm country (where I grew up), it was a well-known fact that livestock attacks by domestic-feral dogs or even coydogs (coyote/domestic-dog crosses) were much worse than a simple coyote attack; the farm would usually suffer a greater loss. Worse, while most predators will go after and prioritize the most vulnerable members of the herd, domestic-feral dogs and even some (nonbreed-specific) domestic dogs will kill indiscriminately.

Finally, in talking about predators, we cannot forget **human predators.** Unfortunately, goat theft is not unusual, and it can be especially prevalent during various religious and ethnic holiday celebrations, when goat, especially young kid, is the main course. As a result, if you are in an area where you feel that this may be a problem for you and your herd (ask the other goat people in your area; they will probably know), it is best to keep your animals locked up before and during these holidays and celebrations.

Not every type of predator will go after every type of goat. Some, like the wild hog or fox, will go after only the small kids. The bear, even though it is considered a predator of goats, does not count goats in its "favorite foods" list. However, should you live in an area with predatory animals (and most of us do), predators would be an unavoidable reality. Thankfully, there are some things that you can do to lessen the risk to your animals.

Predator Prevention

One simple, key act of predator prevention that you can do to protect your animals is to **lock them indoors at night.** Many predators will hunt and feed at night, and animals left outdoors make for easy targets.

When you have one or more does nearing their kidding dates, it will be in your best interest (and theirs) to keep them indoors until a few weeks after kidding. This will allow your kids to become stronger and slightly less vulnerable to predators, as well as give the doe enough time to get its strength back. It will also make safer the possibility of nighttime kidding by eliminating the vulnerability that a kidding doe and/or newborn kid would face if the event occurred outdoors during the night.

Make sure that you have **good, solid fencing.** It should be high enough to keep the goats in and the predators out. It should also be heavy enough so the goats cannot create an escape route through the wire mesh by sticking their heads through the opening and widening it until the wires begin to break and the goats can walk right through.

Goats love to get into mischief when they can, so a strong, dependable fence will save you a lot of headaches going forward.

And, if you think that you may have a problem with other animals burrowing under the fence, bury the bottom of the fence underground or block it off with a dig guard, as was mentioned in Chapter 3. Remember that your goats cannot and will not dig under a fence (although they will crawl out under a fence if they can fit). But, if any other animal digs under—even one that means no harm to the goats—it can become a problem.

Tethering a goat outdoors can make the animal very vulnerable to attack, as it cannot run from its attacker. You should not tether a goat and leave it alone, especially at night. If, for some reason, you do need to tether your goat (maybe you need one to clean out a small, weeded area, which is too small for setting up a temporary fence), then there should be someone attending the goat at all times.

Never leave a tethered goat out overnight. If you are in an area with a lot of predators, leaving a tethered goat out all night is setting the animal up for attack; which, if it happens, will place the blame solely on you. An attack on a tethered animal is preventable simply through responsible ownership.

Guard Animals

Along with locking your animals up at night, there is another precaution that may be taken when pasturing your animals during the day: making use of a **guard animal.**

Before we go further, it should be made clear that a guard animal is usually not a herding animal, and a herding animal is not a guard animal. There are exceptions to this rule, as with my Australian Shepherd, Cheyenne. She is primarily a herding dog; however, she would also defend her charges with her own life if the need arose. Should you be able to find such an animal, it would be one of the most valuable animals on your farm. As stated, though, this would be the exception and not the rule. Do not expect your herding dog to guard, or your guard dog to herd, as you'll be only disappointed if it doesn't happen. And, as always, do your research first.

Guard animals will protect your livestock, sometimes at the cost of their own lives. A good guard animal, like a good herding animal, will be worth its weight in gold. Notice that I say guard *animal.* This is because, contrary to popular opinion, dogs are not the only animals used for guard work: llamas and donkeys are also very popular, and successful, as guard animals. Let's look briefly at the three most popular guard animals just to familiarize ourselves with each.

Dogs

Large guard dogs, also known as **LGDs**, are most likely the first things that come to your mind when you think of guard animals. When raised and trained properly, LGDs will be ruthless against predators, while being as gentle as can be around their livestock charges and human families.

Just like people may work better in one type of a job than another, the same may be said of the LGD. Some may work well on their own, alone with the herd on the back forty somewhere;

others may do better a bit closer to home, having their duties vary. You job is to find the right dog to fit your situation and needs. There are a number of LGDs to choose from. Breeds include:

- Great Pyrenees (aka Pyrenees Mountain Dog)
- Anatolian Shepherd
- Pyrenean Mastiff
- Akbash

This is, of course, only a partial list of LGDs. When you feel that you are ready to purchase your dog, do your homework. Talk with other breeders. If you are unsure about which breed is best for you, have a knowledgeable person work with you on a selection.

Llamas

Llamas make excellent guard animals, and will not only chase off predators, but also will kick and bite them as well. They have even been known to kill an intruder to protect the herd.

Because llamas need to bond with their herds, it is best to have only a single llama in with a small herd. This way, the llama will bond with the herd instead of with another llama. An important note: intact male llamas should *not* be used as guard animals, as they may go after females in the herd that are in heat.

Just like with LGDs, should you decide that a llama will be your guardian of choice, you need to do your homework. And there is one more little perk that comes with owning a llama: you will get to keep all the wool or fiber from its yearly shearing.

Donkeys

If your predator problem is dog, coyote, or the like, then a donkey may be in your future. Like the llama, it will chase, bite, and kick, as well as make a general ruckus when there is trouble about.

Only a jenny (female) or a gelded jack (neutered male) should be used as a guard animal, as an intact jack can become aggressive toward his charges. Again, as with the other guard-animal types, should you decide to use a donkey, do your homework. If you are not sure, work with someone knowledgeable on guard donkeys.

This should give you a good, albeit brief, idea about predators and how to go up against them. From feathers to four legs to two legs, no matter where you live and what you have, predators are a fact of life when you have livestock. However, if you take precautions, use common sense, and do the best that you can to protect your animals (although you may never be 100 percent predator free or even loss free), you will eliminate many problems.

CHAPTER 13

OTHER USES FOR GOATS

Throughout this book, we have discussed raising goats for milk and meat. But, of course, they can have other uses as well, and can be very useful around the farm.

Pets: High on the list is pet use. Goats make great pets, and can be both comical and affectionate. My buck Elvis used to come to the front door for his daily snack, even knocking on the door with his horns if no one saw him or came to the door right away.

Companion animals: Goats naturally like to be with company, and it does not necessarily need to be another goat. This is why they make great companions for other animals. In fact, it is not uncommon for a goat to be a companion for a horse.

Having goats pull carts, for light loads or children, has been popular for a long time. *Three Children with a Goat Cart. Painting by Frans Hals.*

Carts: If trained early, wethers will pull carts. Both harnesses and parts are available that are made especially for goats.

Using goats as pack animals is just another way to make use of your herd.

Pack animals: Wethers also make great pack animals (does can also be used). Although they can't carry what a larger animal can, they can be great company on a day hike, carrying snacks, first-aid kits, their own treats, and other small things that can fit in their packs. There are packs made especially for goats.

Mowing: Retired goats can be excellent to rent out for weeding and mowing. Because they love weeds, saplings, and other "weedy growths" (as discussed earlier in Chapter 4), cemeteries, green businesses, landowners who need to clear out large areas without using chemicals, and many others will use goats to do so, as goats can clean out a large area faster than a human can. Many owners will rent out goats just for "mowing." Some have made quite lucrative businesses out of this.

Hair: Some goats are raised for their hair. Angora Goats are raised specifically for their hair, which, when spun, is known as mohair.

As you raise your goats, you may find your own uses for them around the farm. Many don't even think about the versatility of their goats. But you'll soon see for yourself how much more fun you can have with your goats as you discover the new adventures that you can have with them and how much they'll be able to contribute to the farm besides just milk and meat.

FINAL NOTES

The more you get to know your goats, the more you will see that the goat is a very useful and versatile animal. You'll also see firsthand why they are known as the "clowns of the barnyard."

Goats can contribute quite a bit to their farm: milk, cheese, cream, meat, fertilizer, hair, and more. They take up less space than cattle, and 10 goats can pasture in the same space that would hold only two steers. A goat with a good temperament is an animal that the entire family can handle and work with.

Once you've made that final decision to purchase your first goat or goats, there are thousands of resources available to which you can go for more in-depth information. The "Resources" section of this book provides only a small sampling. And once you get your first goats, don't be afraid to have fun.

Enjoy!

FUN FACTS

- Goats do not have front teeth in their upper jaws.
- It is said that drinking the milk of a goat that eats p ivy will improve your own immunity.
- There are a number of nonfood products that can be m from goat's milk, including soaps, lotions, and creams.
- Goat's milk is easier to digest than cow's milk. Many wh cannot drink cow's milk can drink goat's milk.
- Goats will chew on electrical cords.
- Pygmy goats can be litter-box trained. (I did it. It takes some time and patience, but can be done.)
- If your goat's house has a flat roof, you may find a goat on it.

Goats like to climb, and should have some climbable areas in their pen. However, no climbing area should be placed by a fence, as they will climb up, then jump right over.

…oy enjoying a ride in a goat-pulled cart. Photo by anyjazz65 under the … Commons Attribution License 2.0.

…carts have been used throughout history.

…hough cats are thought of as a witch's familiar, goats and …her animals were said to serve as familiars as well.

…Goats with multiple horns are known as **polycerate**. Some have had up to eight. It is thought to be inherited through genetics.

While it does require slightly more maintenance, polycerate goats do not typically experience difficulty from their extra horns.

RECIPES

Roasted Tomato and Goat Cheese Sandwich

Kim Pezza

6 tomatoes (medium to large), ½ inch slices

Soft goat cheese (your choice of flavor)

Fresh small baguette, cut in half or loaf of French bread cut into thirds or fourths

Olive oil

Balsamic vinegar

Salt, to taste

Fresh cracked pepper, to taste

Sugar, to taste

Garlic powder, to taste

Mix together the salt, cracked pepper, sugar and garlic powder to create a dry rub. Place tomatoes on parchment covered cookie sheet and generously sprinkle rub mix on slices. Bake at 225°F until tomato slices are roasted and dry to the touch, but still quite pliable. This could take 3 or more hours, depending on your oven.

Lightly toast bread. Place roasted tomatoes and crumbled goat cheese on one side of the bread. Place under a broiler until cheese melts slightly. Remove, and top with a little olive oil and balsamic vinegar. Cover with top piece of bread. Serve warm or cold.

Variations:

Sprinkle extra rub on sandwich before serving.

Rub toasted bread with roasted or raw garlic before building sandwich.

Use garlic or roasted garlic oil instead of olive oil.

Goat Cheese Cakes

Wiki Cookbooks

1 hour

11 ounces goat cheese, room temperature
1 (5.2-ounce) package Boursin brand cheese, room temperature
4 ounces Ricotta cheese, room temperature
1 tablespoon butter
1 tablespoon minced garlic
1 tablespoon minced shallot
1 tablespoon minced chives
2 teaspoons minced fresh thyme
2 teaspoons minced fresh basil
2 teaspoons minced fresh tarragon
1 cup flour, for dredging
3 eggs, beaten lightly (to make an egg wash)
About 1 cup bread crumbs
4 tablespoons olive oil

In a mixing bowl, combine goat cheese, Boursin and ricotta cheese, and set aside. Melt butter in a small saucepan over medium-low heat. Add the garlic, shallot, chives, thyme, basil and tarragon. Cook just until soft and fragrant, about 2 minutes.

Fold herbs into cheese mixture. Cover and refrigerate until mixture is firm, at least 30 minutes. When cheese mixture is firm, divide it into 8 equal portions and form each portion into a cake about ¾-inch thick. Dredge each cake in flour, then shake off excess flour.

Dip each cake in egg wash, then into breadcrumbs, covering well on both sides. Set on a wire rack. Heat oil in a large skillet or sauté pan over medium-high heat. Gently add cheese cakes and fry until golden on both sides, 1 to 2 minutes per side. Drain on a wire rack.

Makes 8 cakes.

Red Pepper and Goat Cheese Sauce

Wiki Cookbooks

2 large red bell peppers
1–2 cloves garlic
2 teaspoons olive oil
¼ cup red wine
1 pound pasta
4 ounces goat cheese (bouchon, crottin, or generic chevre)
Basil, to taste
Oregano, to taste
Tabasco sauce, to taste

Skin and finely chop or process the red peppers. Place in saucepan and cover with water. Crush and add garlic (I leave the cloves in; some people may want to fish them out later). Cover and stir in wine and olive oil until simmering.

Add basil, oregano, and Tabasco, stirring occasionally at simmer; allow sauce to begin cooking down. While you're doing this, make the pasta the usual way. Remember that the sauce only needs to take its sweet time to cook down. Cut the rind from the goat cheese; you only want the inner part for this.

As soon as the pasta and sauce are ready, toss them, and the goat cheese, together in your serving dish. You have to do this fast, so the goat cheese will melt through the sauce and over the pasta. Stirring the goat cheese into the sauce could result in burned sauce.

Serves 4.

Jamaican Patty

Wiki Cookbooks

Filling

½ teaspoon Island Spice Scotch Bonnet Pepper Soya Sauce

2 ounces scallion or spring onion

2 pounds ground goat meat

1 teaspoon salt

2 bundles fresh thyme

1 teaspoon paprika

Half a loaf of French bread

Pastry

2 cups flour

½ teaspoon salt

1 cup lard

1 cup cold water

1 teaspoon ground turmeric (or ½ teaspoon Jamaican Curry Powder and ½ teaspoon annatto for coloring)

Filling, for patties

½ cup milk

Filling

Grind scallion and Island Spice Scotch Bonnet Pepper Soya sauce. Add to ground meat with salt and thyme, and cook. While beef is cooking, pour cold water over bread in a saucepan to cover and soak for a few minutes, squeeze dry saving the water.

Pass bread through mincing mill and return the ground bread to the water with thyme and cook until bread is dry. Combine meat and cooked bread. Add paprika and cook for 20 minutes. Remove from heat and allow to cool.

Pastry

Sift flour, turmeric (for coloring), curry powder, salt and knead in lard. Bind with water to form a firm dough and knead for 2 minutes. Roll pastry 1/8 inch thick. Cut pastry round 6 inches. For extra patty crust flakiness, roll out brush with lard, fold over and roll again. Do this about 3 times.

Divide meat filling between patties, brush edges with water. Fold over and seal glaze with milk. Bake on top shelf of oven for about 25 minutes.

African Goat Stew

2 large onions
2 carrots
1½ pounds goat meat
1 clove garlic
1 tablespoon peanut butter
3 tablespoons butter
2 tablespoons tomato puree
½ bay leaf
⅛ teaspoon cloves
⅛ teaspoon ginger
Dash cayenne
Dash salt and white pepper
1 tablespoon freshly squeezed lemon juice
2 cups beef stock
2 tablespoons flour

Peel and dice the onions, and slice the carrots. Finely dice the meat, and crush the garlic. Heat peanut butter to medium heat and sauté the meat and vegetables for a few minutes. Add tomato puree, spices, lemon juice, and stock, and bring to a boil. Reduce heat, cover and let simmer until meat is tender. Mix peanut butter and flour and stir into the stew. Let simmer for a few more minutes and check for seasoning.

Goat Cheese with Spinach and Sun-Dried Tomatoes

Wiki Recipes

¼ pound goat cheese

3 tablespoons olive oil

1 radish, diced

4 tablespoons sun-dried tomatoes

1 cup romaine lettuce

½ cup spinach

1 tablespoon balsamic vinegar

Slice goat cheese, brush with oil and broil until golden. Serve on top of salad with diced radish and sun-dried tomatoes. Drizzle with oil and vinegar.

Pepper Soup

Wiki Recipes

Soup

2 pounds goat meat, diced
1–2 onions, quartered
2–3 hot chili peppers, cleaned and chopped
1 cup water
4 cups meat broth or stock
2 tablespoons ground dried shrimp
1 small bunch fresh mint leaves, chopped
1 tablespoon fresh or dried utazi leaves (or bitterleaf)
Salt and black pepper, to taste

Seasoning

Allspice
Anise pepper
Anise seeds
Cloves
Coriander seeds
Cumin seeds
Dried ginger
Fennel seeds
Tamarind pulp

In a deep pot or Dutch oven, combine meat, onions, chili peppers, and a cup of water. Bring to a boil and cook until meat is done, simmer for 20 to 30 minutes, adding water as necessary to keep pot from becoming dry.

Make a mix from the seasoning mix ingredients. Add the seasoning and the broth or stock (or water) and simmer over low heat for 10 to 20 minutes. Add the dried shrimp, mint leaves, and utazi leaves. Add salt and pepper according to taste.

Simmer until soup is to be served.

Egusi Soup

Wiki Recipes

¾ cup pumpkin seeds, or egusi, usually found in African or tropical food markets

1½ pounds cubed goat meat

½ cup palm oil

1 small onion, chopped

2 habañero peppers

½ cup crayfish

1 tablespoons ogiri

Salt and pepper

Chicken bouillon

Beef stock

1 pound fresh spinach, washed and chopped

Basil (optional)

Place egusi (pumpkin seeds) in a blender and blend for 30 to 40 seconds or until mixture is a powdery paste and set aside. Also blend in the crayfish, ogiri, habañero pepper and half onion, set aside. Mince other half onion into bite-size cubes. Season to taste in a large pan. Heat oil over medium-high heat for 4 to 5 minutes. Add minced onion and sauté for 2 to 3 minutes or until brown, then add the crayfish, pepper and ogiri blend.

Cook for 5 minutes, and then add into already cooked meat with stock. Add blended egusi, stir and reduce heat to low-medium. Add salt and pepper to taste. Cook for 15 to 20 minutes or until meat is tender. Add spinach and continue to simmer 10 minutes more. Optional, add some fresh, chopped basil to increase flavor.

Nigerian Pepper Soup

Wiki Recipes

3 pounds stewing goat meat

1 cup water

1 teaspoon salt

1 medium onion, chopped or ground in blender

2 tablespoons cooking oil

Fresh red tomatoes, chopped or ground in blender (optionally, one 6-ounce can tomato paste)

1 teaspoon dried red pepper, crushed or ground

1 teaspoon curry powder (optional)

2 Maggi cubes (optional)

1 teaspoon thyme leaves (optional)

Cut goat meat into small pieces and place in stewing pot. Combine with water, salt, and onion. Boil until meat is tender. Drain, remove meat from pot and save water in a bowl. Heat oil in stewing pot, add and brown onion. Add meat and ground tomatoes or dilute tomato paste with water and add to meat. Add the remaining ingredients one by one, stirring each as added.

Simmer for 10 minutes and serve.

Goat Cheese-Stuffed Zucchini Blossoms on Tomato Salad

Wiki Recipes

4 ounce fresh, mild, soft goat cheese, at room temperature

2 tablespoons coarsely chopped fresh basil

2 tablespoons coarsely chopped fresh marjoram

8 large fresh zucchini blossoms

3 tablespoons extra-virgin olive oil

2 cups vine-ripened cherry or other small tomatoes, halved or quartered, or two large vine-ripened tomatoes cut in bite-sized pieces

Salt to taste

Preheat oven to 350°F. Lightly oil baking sheet or line with parchment paper, and set aside. Mix cheese with half basil and half marjoram. Form into 8 balls of equal size. Inspect blossoms for insects, and snap off pistils inside flowers with fingertips. Cut stems to about 1 inch. Put ball of cheese inside each blossom, and arrange on baking sheet. Brush blossoms with half the olive oil, and season lightly with salt. Bake stuffed blossoms for 7 to 10 minutes, or until petals collapse onto cheese and sizzle slightly around the edges. Meanwhile, toss tomato halves or slices with remaining olive oil, marjoram and basil and season with salt.

To serve, arrange salad on four plates, and top with warm blossoms.

Home Cheese Making

(from Wikibooks)

Home cheese making has been in practice for thousands of years and comprised nearly all cheese production until the 19th century. While factory cheese production has taken over the majority of the market, many people still make cheese in the traditional fashion.

Milk contains a wide variety of fats and proteins. Some of these are suspended solids and minerals; others are liquids. The process of separating the solids from the liquids is **curdling;** the white solid remainder is known as **curds,** and the greenish liquid remainder **whey.** Cheese is curds in a wide variety of forms. Soft cheeses are little changed from the original curd; they are typically drained but not pressed, and are usually unaged. Semisoft (or semi-hard) cheeses are drained and lightly pressed, and may be aged. Hard cheeses are drained and well pressed, and are almost always aged.

To cause milk to curdle requires a **curdling agent.** There are a wide variety of curdling agents available in nature, both plant and animal based—a quick search of the Internet will show some to you. In practice, only a few are regularly used in cheese making. Vinegar is commonly used in soft cheeses, and also assists in making ricotta; it creates a sticky curd in small flecks. Lemon juice is also used in a few soft cheeses. Tartaric acid is the sharp, lemony curdling agent that makes mascarpone cheese, and creates a very fine sticky curd. For most semisoft and hard cheeses, **rennet** is used. There are three types of rennet in common usage. The most traditional rennet is animal rennet; this is an enzyme taken from the digestive tract of mammals. For vegetarians, more companies are producing a "vegetable" rennet. These are not truly vegetable, but are microbial based.

Color is related to two things: the **natural color** (which is usually a creamy white to pale yellow), and **additives.** The most common color additive is annatto, an extract of the dark red seeds of plants in the Bixacae family, typically grown in South America. Their dark red/orange color dilutes into the typical cheddar-yellow that we're all familiar with. Annatto coloring is generally available in two forms: powder and liquid. The powder can take time to dissolve, and should be added as early in the cheese making process as possible (preferably during pasteurization, as the heat helps it dissolve). The liquid extract is easily mixed in, and can be added at any point before the curdling agent is added.

Traditional food coloring does not work well at all for coloring curds; it tends to remain in the whey instead of the curd. However, adding food coloring to uncolored curds after they have been drained but before they have been pressed leads to an attractive mottled pattern rarely found in commercial cheeses.

The Ten Stages of Cheese Making

1. Pasteurization
2. Cooling (in cold water or snow)
3. Inoculation (in all cheeses that age except surface-ripened cheeses)
4. Curdling (using a curdling agent)
5. Cooking (typically only rennet-based cheeses)
6. Draining
7. Salting (mixed into the curds)
8. Pressing (on most semisoft cheeses and all hard cheeses)
9. Brining (on brined cheeses)
10. Aging (on aged cheeses)

General Rules for Making All Cheeses

- Always pasteurize your milk, even if you bought it from the store.
- When pasteurizing milk, do not forget to stir. If you're forgetful, set a timer. If you forget to stir, the milk will scald; scalded milk should never be used.
- Do your best to avoid contaminating your sample. Try not to put anything that's not clean (including unclean hands) into your cheese.
- Make milk-cheeses in batches of at least two gallons to save time (requires a large pot).

RESOURCES

Websites

Hoegger Supply Company (hoeggerfarmyard.com)

Excellent resource for goat equipment, cheese-making supplies, books, and everything else goat.

Goat-Link (goat-link.com)

Excellent online resource for all types of information about goats.

Purdue University: "Common Diseases and Health Problems in Sheep and Goats"

(www.extension.purdue.edu/extmedia/AS/AS-595-commonDiseases.pdf)

Alabama Cooperative Extension System: "Causes of Infectious Abortions in Goats"

(www.aces.edu/pubs/docs/U/UNP-0079/UNP-0079.pdf)

Ontario Ministry of Agriculture and Food: "Nutrition of the Young Goat: Birth to Breeding"

(www.omafra.gov.on.ca/english/livestock/goat/facts/goatnutrition.htm)

North Carolina State University's Department of Animal Science: "Breeding and Kidding Management in the Goat Herd"

(www.cals.ncsu.edu/an_sci/extension/animal/meatgoat/MGBrdKidd.htm)

American Dairy Goat Association: "Parts of a Dairy Goat"

(www.adga.org)

Nice and simple outline of the dairy goat.

Florida A&M: "Facts About Goats"

(www.famu.edu/cesta/main/assets/File/coop_extension/small%20ruminant/goat%20pubs/Facts%20About%20Goats.pdf)

Basic information on goats, written simply.

Caprine Supply

(www.caprinesupply.com)

Another excellent source of goat supplies, including for cheese-making and working goats.

Working Goats

(www.workinggoats.com)

A great online resource for those interested in the working goat.

“How to Make Goat Harnesses”

(www.ehow.com/how_5757563_make-goat-harnesses.html)

Covers the basics of making a harness for your working goat.

“How to Train a Goat to Pull a Cart”

(www.dummies.com/how-to/content/how-to-train-a-goat-to-pull-a-cart.html)

Tips on teaching your goat to pull a cart.

“How to Make Driving Harnesses for a Goat”

(www.ehow.com/how_5644865_make-driving-harnesses-goat.html)

The Goat Spot

(www.thegoatspot.net)

Interesting and useful online resource.

American Livestock Breeds Conservancy

(www.albc-usa.org)

Organization that focuses on rare-breed and heritage livestock.

New Century Homesteader

(newcenturyhomesteader.blogspot.com)

Workshops and programs. Feel free to contact with questions on goats or any other aspect of backyard and small/no-space farming.

Urban Farm Online

(www.urbanfarmonline.com/urban-livestock/?navm=toplivestock)

Keeping livestock in urban situations. Many articles on goats.

Beginning Cheese Making

http://biology.clc.uc.edu/fankhauser/cheese/cheese_course/cheese_course.htm

A good introduction to cheese making, written by a Professor of Biology at the University of Cincinatti.

Apps

GoatTracker

A free app that helps you keep track of your doe's heat cycle, kidding date, and more.

Livestock Manager

Free application that will assist you in keeping track of your livestock's activities.

Livestock Gestation Tables

Gestation charts and calendar for use with livestock.

Periodicals

Acres USA: The Voice of Eco-Agriculture

(www.acresusa.com)

Excellent magazine for sustainable and organic farming. Lots of articles for the small and backyard farmer.

Countryside & Small Stock Journal
(www.countrysidemag.com)
One of the first in self-sufficiency. Lots of articles on goats.

Grit
(www.grit.com)
Excellent magazine for homesteading, urban farming, and more.

Hobbyfarms.com
(www.hobbyfarms.com)
Publisher of wonderful magazines in all areas of urban farming, backyard farming, and more.

Mother Earth News
(www.motherearthnews.com)
One of the first magazines for those interested in homesteading and self-sufficiency. A variety of articles about goats, their upkeep, and use. Also has articles on building stands.

Books

Bowman, Gail B., and Annette Maze. *Raising Meat Goats for Profit.* Twin Falls, ID: Bowman Communications Press, 1999.

Burch, Monte. T*he Ultimate Guide to Home Butchering: How to Prepare Any Animal or Bird for the Table or Freezer.* New York: Skyhorse Publishing, 2012.

Butchering, Chevon, and Goat Hides. Columbia, MO: *Dairy Goat Journal.*

Cannas, Antonello, Giuseppe Pulina, and Ana Helena Dias Francesconi. *Dairy Goats: Feeding and Nutrition.* Wallingford, UK: CABI, 2008.

Coleby, Pat. *Natural Goat Care.* Austin, TX: Acres U.S.A., 2001.

Dardick, Geeta. *Home Butchering and Meat Preservation.* Blue Ridge Summit, PA: TAB Books, 1986.

Luttmann, Gail. *Raising Milk Goats Successfully.* Charlotte, VT: Williamson Publishing, 1986.

Nutrient Requirements of Goats: Angora, Dairy, and Meat Goats in Temperate and Tropical Countries. Washington, D.C.: National Academy Press, 1981.

Smith, Mary C., and David M. Sherman. *Goat Medicine.* Philadelphia: Lea & Febiger, 1994.

Thedford, Thomas R. *Goat Health Handbook: A Field Guide for Producers with Limited Veterinary Services.* Morrilton, AR: Winrock International, 1983.

Vincent, Barbara. *Farming Meat Goats: Breeding, Production and Marketing.* Collingwood, Australia: Land Links, 2005.

Weems, David B. *Raising Goats: The Backyard Dairy Alternative.* Blue Ridge Summit, PA: TAB Books, 1983.

NOTES

NOTES

NOTES

NOTES

NOTES

NOTES

NOTES

NOTES

NOTES

NOTES

NOTES

NOTES